Will You Be Alive

10 Years from Now?

Will You Be Alive 10 Years from Now?

And Numerous Other Curious Questions in Probability

A collection of not so well-known mathematical mind-benders
(with solutions, with one exception)

PAUL J. NAHIN

PRINCETON UNIVERSITY PRESS
PRINCETON AND OXFORD

press.princeton.edu

Jacket art by Dimitri Karetnikov

Library of Congress Cataloging-in-Publication Data

Nahin, Paul J.
Will you be alive 10 years from now? : and numerous other curious questions
in probability : a collection of not so well-known mathematical mind-benders
(with solutions, with one exception) / Paul J. Nahin.
pages cm
Summary: "What are the chances of a game-show contestant finding a chicken
in a box? Is the Hanukkah dreidel a fair game? Will you be alive ten years from
now? These are just some of the one-of-a-kind probability puzzles that acclaimed
popular math writer Paul Nahin offers in this lively and informative book. Nahin
brings probability to life with colorful and amusing historical anecdotes as well as
an electrifying approach to solving puzzles that illustrates many of the techniques
that mathematicians and scientists use to grapple with probability. He looks at
classic puzzles from the past—from Galileo's dice-tossing problem to a disarming
dice puzzle that would have astonished even Newton—and also includes a dozen
challenge problems for you to tackle yourself, with complete solutions provided in
the back of the book. Nahin then presents twenty-five unusual probability puzzlers
that you aren't likely to find anywhere else, and which range in difficulty from
ones that are easy but clever to others that are technically intricate. Each problem
is accompanied by an entertaining discussion of its background and solution,
and is backed up by theory and computer simulations whenever possible in order
to show how theory and computer experimentation can often work together on
probability questions. All the MATLAB Monte Carlo simulation codes needed to
solve the problems computationally are included in the book. With his characteris-
tic wit, audacity, and insight, Nahin demonstrates why seemingly simple probability
problems can stump even the experts."—Provided by publisher.
Includes bibliographical references and index.
ISBN 978-0-691-15680-4 (hardback)
1. Probabilities—Problems, exercises, etc. I. Title.
QA273.25.N344 2014
519.2—dc23   2013010221

British Library Cataloging-in-Publication Data is available

This book has been composed in ITC New Baskerville
Printed on acid-free paper. ∞
Printed in the United States of America

1 3 5 7 9 10 8 6 4 2

For Patricia Ann

who, after fifty years of marriage, can still surprise me

even more than a good probability problem can!

"If we do everything right, if we do it with

absolute certainty, there's still a 30 percent

chance we're going to get it wrong."

— Joseph R. Biden, vice president of the United States,
during a 2009 political talk in Williamsburg, Virginia,
that showed a shaky understanding of probability is no
roadblock to higher elective office in America

What has hit Los Angeles is a force of nature that is exactly the same as a flood or a fire or an earthquake. It's the natural law of probability. Given four to five million vehicles traveling on those freeways each day, a certain number of accidents will happen. It was mathematically unlikely that one would happen at each of seventeen major interchanges simultaneously on any given day, but the odds caught up with us. Mathematics is a force of nature, too.

From Thomas Perry's very funny 1983 novel *Metzger's Dog*,
in which LA thieves show that the CIA's middle initial
might be claiming just a bit too much

There are few persons, even among the calmest thinkers, who have not occasionally been startled into a vague yet thrilling half-credence in the supernatural, by *coincidences* of so seemingly marvelous a character that, as *mere* coincidences, the intellect has been unable to receive them. . . . Such sentiments are seldom thoroughly stifled unless by reference to the doctrine of chance, or as it is technically termed, the Calculus of Probabilities. Now this Calculus is, in its essence, purely mathematical; and thus we have the anomaly of the most rigidly exact in science applied to the shadow and spirituality of the most intangible in speculation.

The opening words of Edgar Allen Poe's
"The Mystery of Marie Rogêt" (1842/3)

Contents

Preface

Probability problems fascinate just about everyone, despite wildly varying backgrounds that span the sophistication spectrum. Mathematicians love them because probability theory is so beautiful, one of the gems of mathematics. Physicists love them because probability often is the key to answering many of their technical questions. And everybody loves probability problems simply because they can be a lot of fun. A probability problem can simultaneously be so easy to state that anybody with a brain can understand the question, while also being so puzzling that even the experts are stumped (at least for a while).

Here's an example of what I mean by that. Suppose we have an urn with 100 red balls and 100 black balls in it. Suppose further that we select balls, one after the other, *with replacement*—that is, after selecting a ball we put it back in the urn. Then,

(a) What's the probability the first ball is red? The answer is, of course, 1/2.

(b) What's the probability the second ball is red? Ho, hum, you say, it's 1/2, too, of course!

Now, suppose we select balls *without* replacement, that is, after drawing a ball we discard it. Then,

(c) What's the probability the first ball is red? My gosh, you say, this is getting boring—it's clearly still 1/2 because it doesn't matter what we do with first ball after we select it.

(d) What's the probability the second ball is red? Okay, you shout, at last a nonobvious question! *Now*, you surely say, the answer depends on what the first ball was because if the first ball was red, then there would be just 99 red balls in the urn for the second drawing, while if the first ball was black, then there would still be 100 red balls available but only 99 black balls.

This does seem to be more complicated than were the earlier cases. What we *can* write is the following, using what are called *conditional* probabilities:

Prob(second ball is red) =
Prob(second ball is red given that the first ball is red) ×
Prob(first ball is red)
+ Prob(second ball is red given that the first ball is black) ×
Prob(first ball is black)
= (99/199)(100/200) + (100/199)(100/200) = (100/200)
$$(99/199 + 100/199) = (100/200)(199/199) = \frac{1}{2}.$$

Now, if this result doesn't surprise you—it's *still* 1/2—well, all I can say is that I used this example in my classes for decades, and it still makes *me* scratch my head!

Or how about this amazing result. Each of a million men puts his hat into a very large box. Every hat has its owner's name on it. The box is given a good shaking, and then each man, one after the other, randomly draws a hat out of the box. What's the probability that at least one of the men gets his own hat back? Most people would answer with "pretty slim," but in fact the answer is the astonishingly large 0.632! Who would have guessed that?

Probability has much to offer to analysts who fall outside the usual techno-types, too. For example, students of the social implications of new laws can often use the mathematics of chance to explore what might happen in the future. That claim may seem a bit murky, so let me provide a specific example of what I mean. The topic of what to do about undocumented immigrants has been simmering for years in America, and exploded onto center-stage during the intense political debates of the 2012 presidential election year. One proposed law would allow police officers to stop any person, any time, and demand to see citizenship identification papers. There are certainly all sorts of issues one might have with such a law, but let's concentrate on one in particular—how intrusive, that is, how *inconvenient* would such a law be? We can model this question as a traditional probability problem, as follows.

Imagine America as a huge urn and its population as balls: *b* black balls for legal residents and *r* red balls for undocumented residents.

A police stop is equivalent to drawing a ball at random from the urn. A black ball (legal) is returned to the urn, while a red ball (undocumented) is permanently discarded (that is, deported). Our inconvenience question is then just this: on average, how many times is each black ball drawn (a legal resident stopped by police) before 50%, or 90%, or 99% of the red balls (undocumented residents) have been removed? If we can answer this question (and we will later in this book), then the specific answers are actually not the central point. What *is* the central point is that a probability analysis will have given us some numbers to discuss, and not just emotionally charged words for advocates on each side to throw at each other like rocks.

I taught probability theory and its applications to electrical engineering students at both the undergraduate and graduate level (at the University of New Hampshire and the University of Virginia) for nearly thirty years. I sincerely hope, in my retirement, there are in the world still students of mine who, if asked, would say "Yes, Professor Nahin did teach me some stuff that I have found, now and then, to be useful." But of course, the learning process isn't just a one-way process. As any honest teacher will admit, it isn't only the students who learn new things in school, and I was no exception. Two very important lessons I learned during those three decades of lecturing and chalking mathematics on blackboards were:

(1) "Obvious" results are boring; students are interested in (and pay attention to) calculations that result in nonintuitive and/or surprising (almost "seemingly impossible") results.

(2) While theoretical proofs are nice, and indeed to be hoped for, students are inherently skeptical. They love to ask, "But how can you be *sure* you haven't overlooked something, or haven't made some subtle error of reasoning?"

There are numerous examples of both cases in modern books on probability theory and its applications. To illustrate case (1), for example, the well-known birthday problem is virtually certain to appear in any undergraduate probability book pulled at random from the stacks of a college library. The problem is easy to state. Assume birthdays are randomly distributed uniformly over all the days in the year. Then, if students are asked how many people would have to gather together to achieve a probability of at least 0.5 that two (or more) of them have the same birthday (month and day), most guess a "large" number such as 183 (the first integer greater than half of 365). Literally *everybody* is astonished at how small the actual value is (just 23). If we raise the probability to 0.99, the number of people required increases only to 55, still an astonishingly small value. A twist on the birthday problem, with an equally surprising answer, is to ask how many people you'd have to ask to have a probability of at least 1/2 to find somebody with *your* birthday. Now the answer *is* large, even greater than 183 (it's 253).

I have often used the first birthday question in large introductory classes (say, 40 to 50 students) by having the students, one after the other, call out their birthday; to hear the collective gasp of astonishment when, almost always, there would be a match (often very quickly, too) was a lot of fun for all. (Once we had a triple match!) The students were intrigued by the possibility of actually being able to calculate such an amazing result, and thereafter paid more attention to what I said in class—or at least they usually did until the end of that lecture! So the two birthday problems are undeniably great problems to include in any beginning course on probability theory. But that's the very reason they are not analyzed here: they are simply too well known to be included in this book. To get into this book, a case (1) problem has to be both amazing and not well known. For this reason, too, the wonderful Buffon needle problem for "experimentally" determining the value of π is also a no-show here.

In some respects, this book resembles the 1965 classic by the late Harvard mathematician Frederick Mosteller, *Fifty Challenging Problems in Probability with Solutions.* Nearly half a century has passed since its appearance, and most of the problems in that book have become regulars in textbooks. They are all great problems, but they are no longer "curiously odd and strange" problems. The million-hats-in-a-box puzzle I mentioned earlier, for example, is discussed in Mosteller's book, but even when he wrote, the problem was already centuries old, dating back to the early 1700s.

I've already used the balls-in-urns imagery twice in this preface, so here's another fun example of what I do think amazing, in the form of another balls-in-urns puzzle for you to mull over as you finish reading this essay. I'll give you the answer at the end. Suppose you have 10 white balls and 10 black balls that you can distribute any way you like between two urns. After you've done that, you give a friend a fair coin and he selects an urn at random (tails he picks one, heads he picks the other). From that urn he draws a ball at random. How should you have distributed the balls to maximize the probability your friend draws a white ball? If, for example, you had put all the white balls in one urn and all the black balls in the other urn, then the probability of drawing a white ball would be $1/2(10/10) + 1/2(0/10) = 1/2$. On the other hand, if you had put 5 black balls and 5 white balls in each urn, the probability of drawing a white ball would still be $1/2$ because $1/2(5/10) + 1/2(5/10) = 1/2$. You can do a lot better than this, however, with a different distribution, and I think you'll be amazed at how much better you can do. And once you've found the answer, can you then see what distribution of the balls *minimizes* the probability that your friend draws a white ball?

Here's another probability puzzler with an amazing answer, made doubly amazing because no math is required, just logical reasoning. Suppose there are 100 people at an airport waiting to board a plane. All have a boarding pass, with an assigned seat. It's a full flight; there are exactly 100 seats on the plane. The first person to board is a free spirit and simply takes a seat at random. It might in fact *be* his assigned seat, but if it is, it's just by chance. Thereafter, however, all other persons boarding, one after the other and being more willing to follow the rules, each do go to their assigned seat and take it unless it is already occupied. In that case they too simply take an unoccupied seat at random. What is the probability that the last person to board still gets his assigned seat? The answer is at the end of this preface.

What counts as amazing does not include "surprising" results arrived at by erroneous reasoning. Here are two examples of what I mean by that perhaps curious comment. For my first example, in her July 31, 2011, "Ask Marilyn" column in *Parade Magazine*, the famous (some mathematicians might substitute the word *infamous*) puzzler Marilyn vos Savant asked her readers the following probability question:

Say you plan to roll a [fair] die 20 times. Which of these results is more likely:

(a) 11111111111111111111
(b) 66234441536125563152?

The answer given by vos Savant was,

> In theory, the results are equally likely. Both specify the number that must appear each time the die is rolled. . . . Each number (1 through 6) has the same chance of landing face-up [that is, 1/6]. But let's say you tossed a die out of my view and then said that the results were one of the above. Which series is more likely to be the one you threw? Because the roll has already occurred, the answer is (b). It's far more likely that the roll produced a mixed bunch of numbers than a series of 1's.

Is vos Savant's answer to the original question correct? Is vos Savant's answer to her extended version of the original question correct? Think about this for now. I'll give you the correct reasoning in just a bit.

For a second vos Savant example of really bad mathematical reasoning, consider her Christmas 2011 column in *Parade Magazine*, in which she printed the following letter from a reader:

> I manage a drug-testing program for an organization with 400 employees. Every three months, a random-number generator selects 100 names for testing. Afterward, these names go back into the selection pool. Obviously, the probability of an employee being chosen in one quarter is 25 percent.
>
> But what's the likelihood of being chosen over the course of a year?

Marilyn's answer was,

> The probability remains 25 percent, despite the repeated testing. One might think that as the number of tests grows, the likelihood of being tested increases, but as long as the size of the pool remains the same, so does the probability. Goes against your intuition, doesn't it?

Yes, it certainly does, perhaps because her comments are so astonishingly, indeed breathtakingly, wrong. Again, think about her correspondent's question while you read, and I'll give you the correct solution in just a bit.

Now, what about case (2) that I mentioned earlier? What I did when teaching, and have done in two previous probability books published by Princeton University Press (*Duelling Idiots and Other Probability Puzzlers* and *Digital Dice*), was endorse the use of computer simulation to check theoretical results. If a computer simulation of a random process agrees (to some acceptable degree) with a theoretical result, then I think one's confidence in both approaches is enhanced. Such an agreement doesn't, of course, prove that either result is correct, but surely one would then have to believe that a remarkable coincidence had instead occurred.

So computer simulations play a big role in this book, but I want to emphasize that it is a book on analysis and not a programming book. I use MATLAB®, but I've written all the computer codes in such a low-level way that you should have little trouble converting any of them to your favorite language. MATLAB® experts may have to bite their tongues, however, at my avoidance of the advantages of the powerful vector/matrix structure of the language. That structure does indeed allow spectacular decreases in computational times, when compared to what can be done with *for*, *if/else*, and *while* loops. I've made extensive use of such loops, however, precisely because I don't want to restrict this book's codes to being MATLAB®-only codes.

All Monte Carlo codes, of course, use a random number generator, a feature available in any modern scientific programming language (every time you call the generator, you get back a number from a distribution uniformly spread from 0 to 1), and MATLAB® is no exception. For more on how random number generators actually work—something you really don't need to know to use them—see *Duelling Idiots* (pp. 175–197). You may find the brief Technical Note at the end of this book helpful, too.

There is an interesting feature to doing computer simulations that I have noticed, after decades of writing computer codes in different languages to implement them. Problems that are hard to do theoretically may require only an easy simulation; the converse may also be true, that is, a problem easy to analyze theoretically may require a complicated simulation code. In the introduction you'll find examples of the use of a computer simulation to check a theoretical derivation, as well as an amusing problem (until recently unsolved) for which I think a computer simulation would be very difficult to write (I don't even try, but don't let that stop *you*).

In the rest of this book I'll assume that you have had some encounters with probability arguments in the past. But I'll not assume it was necessarily the recent past! For example, I'll not bother with defining a binomial coefficient, or with extensive explanations either of Bayes' theorem from conditional probability or of why the expected value of a discrete random variable X that takes its possible values from the non-negative integers is given by

$$E[X] = \sum_{j=0}^{\infty} j \operatorname{Prob}(X = j).$$

On the other hand, when a discussion includes the concepts of the joint probability density and distribution function for the continuous random variables X and Y, $f_{X,Y}(x,y)$ and $F_{X,Y}(x,y)$, respectively, I will remind you that

$$f_{X,Y}(x,y) = \frac{\partial^2 F_{X,Y}(x,y)}{\partial x \partial y}.$$

What to assume you already know and what you might need a reminder of is a bit of a guess on my part, and I can only hope I've been right more often than wrong. In any case, where I've guessed wrong, I assume you can look things up on your own.

Now, what about that die-rolling question posed by Marilyn vos Savant? In fact, vos Savant was correct with her first answer but wrong with her second. Indeed, her proper explanation for the first part is precisely why she is wrong in the second part. Both sequences are still equally likely. I finally understood the origin of her confusion when I read her follow-up column in *Parade Magazine* on October 23, 2011. In that column she printed a letter from a reader who correctly claimed that vos Savant was wrong. Vos Savant continued to claim she was right, and then presented a meaningless, almost incoherent defense; at one point she argued that she had actually rolled, once, a die 20 times and, since she did get a jumbled sequence of numbers, that proved her right. Her last line was the clue for me to where she had stumbled: after rolling a die 20 times the result is, she stated, "far more likely to be . . . a jumble of numbers."

And, of course, that is correct, because there are a lot of jumbled sequences that could occur but only one sequence of all 1s. But that's all beside the point because that wasn't the original question. The original question asked about comparing the probability of the sequence of

20 1s with the probability of *one particular* jumbled sequence of 20 digits, namely, 66234441536125563152. Rolling the die "out of her view" is simply irrelevant. And after rereading her comments in the original July 31 column, I realized she had already made the "mixed bunch of numbers" assertion. When you present a problem, ask for a solution, and then correctly answer a *different* question, that doesn't give you the right to claim everybody else is wrong about the original question.

What about vos Savant's drug testing question? Although I suspected it was going to be a waste of my time, I couldn't resist sending the following e-mail to her via *Parade Magazine*.

Dear Marilyn,

*If the probability of being selected for a quarterly drug test is 0.25, then the probability of **not** being selected is 0.75. Thus, the probability of **not** being selected four straight times (four quarters in a year) is $(0.75)^4 = 0.3164$.*

That means the probability of being selected at least once is $1 - 0.3164 = 0.6836$, not the 0.25 you stated.

Best, Paul

Paul J. Nahin

Professor Emeritus of Electrical Engineering/University of New Hampshire

Not surprisingly, I never received a reply, but in her January 22, 2012, column she finally admitted her mistake, writing, "My neurons must have been napping." There then followed some hand-waving comments that seemed to imply she was thinking of the constant probability of being selected for any given quarterly test, but why should that "go against your intuition"? It is, in fact, quite clear that this trivial interpretation is not what she had in mind when she wrote her original answer.

This all just goes to show that, when it comes to probability questions, even a person with the "highest recorded IQ" can tumble headfirst into a tar pit of nonsense, even with as elementary a question as the die-tossing problem, thus illustrating for us all the value and importance of due diligence. In the introduction you'll see how even the super-genius of Isaac Newton took a tumble over a probability problem involving dice. As the great Victorian mathematician Augustus De Morgan (1806–1871) once wrote, "Everyone makes errors in probabilities, at times, and big ones."

One tumble that De Morgan might have had in mind was mentioned by one of his fellow countrymen, Isaac Todhunter (1820–1884), in his 1865 classic, *A History of the Mathematical Theory of Probability*. There, commenting on the faulty reasoning by the French mathematician Jean Le Rond d'Alembert (1717–1783) in analyzing a coin-flipping problem, Todhunter wrote, "Any person who wishes to see all that a great mathematician could produce on the wrong side of a question should consult [D'Alembert's 1754 essay]." So, even mathematicians make mistakes in their arithmetic (see Problem 24 for a discussion of D'Alembert's goof). I do suspect, however, that the ghosts of De Morgan and Todhunter would swallow *very* hard at Vos Savant's prize-winning goofs!

A final comment on the nature of the problems you'll find in this book. With few exceptions, they are not simply "fun puzzles for a party," as are the birthday problems. The level of difficulty varies over a pretty wide spectrum. Some are straightforward applications of not much more than high school algebra (but with surprising conclusions), while others—not to be overly dramatic—are mathematically very involved. These latter problems are more properly thought of as "puzzles at the edge of the doable." That, I think, is their special charm, that they have extreme complexity that stays just within the borderline reach of undergraduate mathematics. Now, lest the book be thought all serious business, a few of the puzzles perhaps are more in the party category than not, in particular the penultimate one on chickens in boxes that I've taken from an old Marilyn vos Savant column (a puzzle question with which she challenged readers but, as far as I know, she never answered). So, I think that both partygoers and dedicated analysts will find stimulating problems on the pages that follow.

Solution to the Ball-Distribution Puzzle

Put b black balls and w white balls into one urn, and the remaining $10 - b$ black balls and $10 - w$ white balls into the other urn. The probability of drawing a white ball from a randomly selected (using a fair coin) urn is then given by

$$\left(\frac{w}{w+b}\right)\frac{1}{2} + \left(\frac{10-w}{20-b-w}\right)\frac{1}{2}.$$

There are just enough possibilities to check to make it a pain to do them all by hand, but for a computer code it's duck soup. And that's what I did, letting the code run through all the possibilities of w and b, each independently varying from 0 to 10 (a total of 121 combinations); the code found that the maximum probability occurs when $b=0$ and $w=1$. That is, one urn has just a single white ball, while the other urn has all 10 black balls and the remaining 9 white balls. This gives the probability for drawing a white ball as

$$(1)\left(\frac{1}{2}\right) + \left(\frac{9}{19}\right)\left(\frac{1}{2}\right) = \frac{14}{19} = 0.7368, \tag{1}$$

a value significantly greater than 0.5. Notice that minimizing the probability of drawing a white ball is equivalent to maximizing the probability of drawing a black ball. By symmetry with the original question we know that $b=1$ and $w=0$ is the distribution that maximizes the probability of drawing a black ball (which is, of course, again 14/19). But that means the minimum probability for drawing a white ball must be $1 - 14/19 = 5/19 = 0.2632$.

Solution to the Airplane-Seating Puzzle

The answer is 1/2, and not just for 100 people but for any number of people. The key to understanding this problem is realizing that the last person to board will not end up with just any seat out of the 100 (or whatever the number may be) seats on the plane. Instead, he will get either his assigned seat or the seat assigned to the first person who boards. This claim initially strikes most people as surprising, but here's how it works. If the first person to board does take (by chance) his assigned seat then the last person to board definitely gets his assigned seat because all the other people (following the rules) go to their assigned seats. If, on the other hand, the first person to board takes, let's say, seat 10, then the next persons to board with seats 2 through 9 take their seats, too. The next person, the 10th person, then has three possibilities: (1) he takes the last person's seat, (2) he takes the first person's seat, or (3) he takes some other still empty seat. If he takes the first person's seat, then all the other people (including the last person) get their assigned seat. If he takes the last person's seat, then all the other people get their assigned seat except for the last person who has to take the lone remaining seat, the seat assigned to the first person. Finally, if he takes some other seat, say seat 27, then we are back in the same situation with persons 11 to 26 (who simply go to their assigned seats) and the fellow with seat 27 playing the same role as did the person with seat 10. The end result is that the last person to board ends up either with his assigned seat or with the assigned seat of the first person. Every time a person who is boarding can't take his assigned seat because it's occupied, he chooses randomly among all the remaining seats, which includes the seats of the first and the last persons to board. So, at all times there is nothing that distinguishes these two particular seats, and thus they have the same probability of being the seat of the last person to board. And since they are the *only* possible seats for the last person to board, each has probability 1/2.

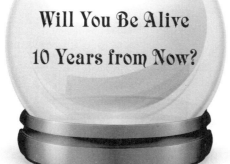

Introduction

CLASSIC PUZZLES FROM THE PAST

I.1 A GAMBLING PUZZLE OF GOMBAUD AND PASCAL

The birth of probability theory is generally dated by historians of science to 1654. That was the year the French writer and philosopher Antoine Gombaud (1607–1684), better known today by his literary name, Chevalier de Méré, presented the French mathematician Blaise de Pascal (1623–1662) with some questions concerning games of chance.[1] This is a bit of an artificial milestone as it refers to the start of an *analytical* approach. Probabilistic *thinking* can be traced back to well before 1654. A hundred years earlier, for example, the Italian Gerolamo Cardano (1501–1575) had pondered probability, and wrote his work up in a 15-page paper, *Liber de ludo aleae* (The Book on Games of Chance). It was written in 1564 but not published until many years later (1663), a decade after Gombaud and Pascal. Furthermore, Cardano thought random behavior was the result of the influence of external, supernatural forces, what he mysteriously called the "authority of the Prince." (This is not the way to a true, mathematical understanding of probability!) One might even make a (weak) case for an even earlier interest in probability, perhaps going all the way back to ancient times in the Holy Land.[2]

For now, however, let's be traditional and go with the Gombaud-Pascal interaction as "the beginning." In particular, Gombaud's befuddlement over two particular games involving dice gives us an interesting peek into how, just a few hundred years ago, the theory of probabilities was still very new. Starting with a fair die, Gombaud correctly understood that on a single toss of a fair die he would observe a 6 with probability 1/6. He also knew that, given two fair dice, a single toss of the pair would show a double 6 with probability 1/36. From that starting point, Gombaud was puzzled when he came to compare two then popular dice games. For the first game one tosses a single die n_1 times; the question for this game is: what should n_1 be to first make the probability of seeing at least one 6 greater than 1/2? That is, what should n_1 be to make an "even money" bet on seeing at least one 6 an *advantageous* bet? For the second game, one tosses a pair of dice n_2 times, and the question is, what should n_2 be to first make the probability of seeing at least one double 6 greater than 1/2?

Gombaud believed in strict proportionality, and so he asserted that it should be true that

$$\frac{n_1}{6} = \frac{n_2}{36}, \tag{I1.1}$$

since in the first game there are six ways for a die to fall, and 36 ways for a pair of dice to fall in the second game. In fact, however, very careful observation of the outcomes of a large number of both games had convinced Gombaud (correctly) that strict proportionality did not hold. In his letter to Pascal, Gombaud had strongly expressed his outrage at the failure of proportionality; as Pascal wrote (in a letter dated July 29, 1654) to his fellow Frenchman, the now world-famous Pierre de Fermat (1601–1665), Gombaud thought this failure "was a great scandal which made him proclaim loudly that . . . Arithmetic belied herself." Pascal was able to properly *calculate* the values of n_1 and n_2, and that is why historians have focused on the 1654 date as the beginning of probability theory.

To calculate the values of n_1 and n_2, write P_1 and P_2, respectively, as the probabilities of first seeing (with a probability greater than 1/2) at least one 6 in n_1 tosses of a fair die, and of seeing at least one double 6 in n_2 tosses of two such fair dice. For P_1, Pascal observed that, in n_1 tosses of a die there are 6^{n_1} ways it can land, with 5^{n_1} of them having not a single 6. So, $6^{n_1} - 5^{n_1}$ is the number of ways *at least one* of the tosses will be a 6. We therefore have

$$P_1 = \frac{6^{n_1} - 5^{n_1}}{6^{n_1}} = 1 - \left(\frac{5}{6}\right)^{n_1} = \begin{matrix} 0.4212\ldots \text{if } n_1 = 3, \\ 0.5177\ldots \text{if } n_1 = 4 \end{matrix}, \tag{I1.2}$$

and so $n_1 = 4$.

A modern mathematician would argue his or her way to the same result using just slightly different language, observing first that the event "no 6 in n_1 tosses" and the event "at least one 6 in n_1 tosses" are *complementary* events, that is, the two events are *mutually exclusive* (if one occurs when the die is tossed, then the other doesn't), and also that the two events are *inclusive* (one of the two events *must* occur). Since the probability of throwing a 6 on a single toss is 1/6, then the probability of not getting a 6 is 5/6, and so, again,

$$P_1 = 1 - \left(\frac{5}{6}\right)^{n_1}.$$

In the same way,

$$P_2 = 1 - \left(\frac{35}{36}\right)^{n_2} = \begin{array}{l} 0.4914\ldots \text{ if } n_2 = 24 \\ 0.5055\ldots \text{ if } n_2 = 25 \end{array}, \tag{I1.3}$$

and so $n_2 = 25$, and not the 24 that Gombaud thought it should be from (I1.1).

Our modern mathematician would explain Pascal's approach as the calculation of the ratio of "favorable" possibilities (that is, the events of either "at least one 6" or "at least one double 6") to the total number of possibilities. The total number of possibilities is called the *sample space* of an *experiment* ("tossing a die n_1 times" for P_1 and "tossing two dice n_2 times" for P_2). The tacit assumption behind taking a ratio is that all of the possibilities in the sample space are equally likely.[3]

I.2   GALILEO'S DICE PROBLEM

Years before Pascal answered Gombaud's puzzle, the Italian mathematical physicist Galileo Galilei (1564–1642), better known today for dropping balls off the top of the Leaning Tower of Pisa, successfully answered a different dice-tossing question from a gambling friend. The friend was puzzled by the following fact: even though with a single toss of three dice the totals of 9 and 10 can each be achieved with the same number (six) of different *sums* (for example, $9 = 4 + 3 + 2$ and $10 = 4 + 4 + 2$), experience shows that 10 is (slightly) more likely to be the result than is 9. Galileo's correct answer was that of the $6^3 = 216$ possible ways for the three dice to fall, slightly more of those ways give a total sum of 10 than the number of ways that give a total sum of 9.

Galileo answered his friend's question with a laborious hand enumeration of the 216 possibilities, but for us it is far easier to write a computer program to run through all the grubby arithmetic. A MATLAB® code that generates all the possible ways three dice can fall on a single toss, and that keeps track of how many ways each possible total is generated, is shown below as the code **galileo.m**. When run, $total(9) = 25$ while $total(10) = 27$. (The total of 11 also appears 27 times, too.)

```
galileo.m
total=zeros(1,18);
for i=1:6
    for j=1:6
        for k=1:6
            s=i+j+k;
            total(s)=total(s)+1;
        end
    end
end
total
```

One attractive feature of this code is the ease with which it can be modified to examine any other number of dice. For example, in the case of four dice (with $6^4 = 1,296$ possibilities, a lot to do by hand), simply add another *for/end* loop for the variable l (that is, change the command for s to $s=i+j+k+l$, and change the first line to $total=zeros(1,24)$. The result is that the most likely total for a single toss of four dice is 14 (it can occur in 146 ways).

I.3   ANOTHER GOMBAUD-PASCAL PUZZLE

Unlike Galileo's problem, or even his "single 6" and "double 6" questions, Gombaud posed another question to Pascal that would have enormous consequences for the history of mathematics. It's called the problem of points. Imagine two players, A and B, who continually flip a fair coin. When heads shows, A wins a point, while when tails shows, B gets a point. Each has put an equal amount of money into a pot, with the entire pot to go to the first one to win n points. At some instant during this process, before either player has won, they decide to call it quits and to split the pot "fairly" based on their present point totals. (In a 1603 Italian text on arithmetic, for example, A and B are two boys who play a series of games involving a ball; before arriving at an overall winner, they lose the ball!) If A has a points and B has b points, how should they split the pot?

This is a very old problem. Its history can be traced back, in Italy, to at least 1380, while Ore (see note 1) conjectured it has an even more ancient, Arabic origin. Before (and after, too!) Pascal finally solved this problem, it was rightfully considered to be one of very great difficulty.[4] The Italian mathematician Niccolo Fontana (1500–1557), for example, better known in mathematical history as Tartaglia ("the stammerer") and famous for discovering how to solve cubic equations, declared in his 1556 book, *General Trattato* (*General Treatise on Number and Measure*), the problem of points to be a question that "is judicial rather than mathematical, so that whatever way the division is made there will be cause for litigation." Tartaglia was a most able analyst, but those words sound (to me) just a bit like a justification for not being able to solve the problem!

Pondering the problem of points is what led Pascal into his correspondence with Fermat, which in turn developed into a deeper consideration by both men of the mathematics of chance. In particular, in 1656 Pascal posed a new problem to Fermat, one so difficult he wondered if Fermat could actually solve it. This was the earliest version of the now famous *gambler's ruin* problem, which can be found in virtually all modern probability texts. Imagine our two players, A and B, again playing a series of identical games. A wins any one of these games with probability p (and so, of course, B wins any individual game with probability $q = 1 - p$). A "game" could be, for example, as simple as flipping a coin, once, with heads showing with probability p (with A winning with heads). Between the two players there is a total, fixed wealth of $a + b$ dollars; at the start of the series of games A has a dollars and B has b dollars. The loser of an individual game gives one dollar to the winner. The games continue until one of the players has lost all of his money (or, as mathematicians put it, is ruined). The original, fundamental question associated with the gambler's ruin problem was that of calculating the probability A is ruined, which of course is one minus the probability that B is ruined, since one of them is certainly ruined in the end.

That calculation was first done by Pascal, using his method of expected values and the mathematics of difference equations,[5] an approach discussed in one of my earlier books so I won't repeat it here.[6] (We'll later encounter difference equations—one of them nonlinear—in the last sections of this introduction, in two entirely different probability problems.) What I will show you, in the next section, is a different, far more ingenious derivation of the ruin probabilities. The work of Pascal and Fermat eventually expanded to include that of the Dutch mathematical physicist Christiaan Huygens (1629–1695), who added the new twist of game duration, that is, the probability one of the players is ruined within n games. All of this work by these three men remained known only to them until Huygens published his "*De ratiociniis in ludo aleae*" ("On calculations in games of chance") in 1657.[7]

I.4　GAMBLER'S RUIN AND DE MOIVRE

In this section I'll show how the French-born English mathematician Abraham de Moivre (1667–1754) derived the ruin probabilities. We can then use those probabilities to find the average number of individual games played until one of the players is ruined. De Moivre's derivation is elegant and incredibly clever,[8] but it is not the one typically presented in modern probability textbooks.[9] After reading through it, however, I predict you'll fully agree with his well-known friend, Isaac Newton (1642–1727), who, when asked about some mathematical question, would often reply, "Go to Mr. de Moivre, he knows these things better than I do." Pretty impressive praise, indeed, coming from one of the inventors of calculus!

To start, suppose we already have the ruin probabilities in hand. That is, we have

$$P_A = \text{probability that A is ruined}$$

and

$$P_B = \text{probability that B is ruined,}$$

where, of course,

$$P_A + P_B = 1. \tag{I4.1}$$

On any individual game, A "expects" (here we see the influence of Pascal) to win one dollar with probability p and "expects" to lose one dollar with probability q. His expected gain *per game* is therefore given by $p(+1) + q(-1) = p - q$. A's expected *total* gain is given by $bP_B - aP_A$ because A wins *all* of B's starting money (b dollars) if B is ruined and loses *all* his own starting money (a dollars) if he himself is ruined. If the average number of games played until somebody is ruined occurs is m, then m times the expected gain per game for A must equal his total expected gain, and so

$$m(p - q) = bP_B - aP_A,$$

or, the average number of games played until ruin occurs is

$$m = \frac{bP_B - aP_A}{p - q}. \tag{I4.2}$$

My gosh, we've suddenly solved the problem of the average duration of the gambler's ruin problem! But of course, (I4.2) isn't of much use until we know the ruin probabilities, and so we can't put off their calculation any longer.

Forget, for now, money being valued in dollars. Let's simply say that A starts with a chips and that B starts with b chips. When it suits us, we can say that a chip is valued at a dollar, but for now a chip is simply a round piece of plastic or cardboard. Imagine that A and B each stack their chips in a pile and that, when one loses a game, they take the top chip of their pile and place it on the top of the other's pile. Imagine further that A assigns a value of q/p to his bottom chip, of $(q/p)^2$ to the next chip up, of $(q/p)^3$ to the next chip up, and so on, so that at the start of play his top chip has value $(q/p)^a$. And finally, imagine that B starts the valuation of his chips from where A left off, but now going top down. That is, before the start of play the top chip in B's pile has value $(q/p)^{a+1}$, the chip immediately below the top chip has value $(q/p)^{a+2}$, and so on down to the bottom chip, which has value $(q/p)^{a+b}$. These chip values remain associated with each chip, no matter what position or which pile they happen to be in as play continues.

In the first game played, A wins B's top chip with probability p, and loses his own top chip with probability q. A's expected gain from the first game is then

$$p\left(\frac{q}{p}\right)^{a+1} - q\left(\frac{q}{p}\right)^{a} = p\left(\frac{q}{p}\right)\left(\frac{q}{p}\right)^{a} - q\left(\frac{q}{p}\right)^{a} = 0.$$

Since A's gain is B's loss (and vice versa) in any individual game, B's expected gain from the first game is also zero (zero having the nice property of being its own negative). Now, before we continue, we need an explanation for this perhaps surprising result. How, you may wonder, can A's expected gain from the first game be zero when, just moments before, we calculated A's expected gain for any game (including the first one) to be $p - q$? Since $p - q = 0$ only for the single, special case of $p = q = 1/2$, don't we have a conflict?

Well, no. There is actually no contradiction here because our first calculation took the value of every chip to be the same, one dollar, while in our second calculation the chips have variable value. In an actual playing of gambler's ruin we would of course use the first valuation, and $p - q$ is indeed A's expected gain in the real world. We use the second valuation of the chips to calculate the *ruin probabilities* in the mathematical world, however, precisely because of the zero expectation for A's gain (and not only for the first game but on all subsequent games, too, as you'll soon see). The actual gain does, of course, depend on how we value the chips, but it should be clear that the probabilities of them all ending up with one player or the other are independent of how we value the chips. So, we'll use a particular chip valuation that is particularly helpful for calculating probabilities. It is that zero result for the expected gain per game, *true only for de Moivre's particular valuation for the chips*, that is the genius of his derivation. Okay, back to business.

A and B start the second game either (if A won the first game) with A having $a+1$ chips and B having $b-1$ chips, or (if A lost the first game) with A having $a-1$ chips and B having $b+1$ chips. In the first case A's top chip has value $(q/p)^{a+1}$ and B's top chip has value $(q/p)^{a+2}$, and in the second case A's top chip has value $(q/p)^{a-1}$ and B's top chip has value $(q/p)^{a}$. Notice carefully that in both cases the chip that B has at risk in the second game is worth q/p times the chip A has at risk in the first game. Thus, A's expected gain in the second game is again zero (as is B's). Indeed, if you continue this line of reasoning on to the third game and beyond, you should see that the expected gain for both A and B, for every individual game, is zero.

This is all pretty clever, but now comes the real heart of de Moivre's derivation. Since A expects to gain zero at every game, then A's total expected gain, no matter how many games the entire gambler's ruin process may require, is zero, since zero expected gain per game times any number of games played is zero. (Remember, this wonderful property is because of De Moivre's particular valuation of the chips.) This conclusion, of zero total gain, is the key to solving for the ruin probabilities P_A and P_B. That's because we can also write A's expected total gain in terms of P_A and P_B, and then set that equal to zero. Here's how to do that.

A's total expected gain is: all of B's starting chips if B is ruined, minus all of A's own starting chips if he himself is ruined. That is, A's total expected gain is

$$\left[\left(\frac{q}{p}\right)^{a+1} + \left(\frac{q}{p}\right)^{a+2} + \ldots + \left(\frac{q}{p}\right)^{a+b}\right] P_B$$
$$-\left[\left(\frac{q}{p}\right) + \left(\frac{q}{p}\right)^2 + \ldots + \left(\frac{q}{p}\right)^a\right] P_A = 0.$$

Summing the two geometric series in the brackets and simplifying, we get (we do have to assume that $p \neq q$, that is, $q/p \neq 1$, to avoid division by zero issues)

$$\frac{P_A}{P_B} = \frac{(q/p)^a - (q/p)^{a+b}}{1 - (q/p)^a}. \tag{I4.3}$$

Combining (I4.3) with (I4.1), a little algebra then gives us the ruin probabilities:

$$P_A = \frac{1-(p/q)^b}{1-(p/q)^{a+b}}, \quad \frac{p}{q} \neq 1, \tag{I4.4}$$

and

$$P_B = \frac{1-(q/p)^a}{1-(q/p)^{a+b}}, \quad \frac{q}{p} \neq 1. \tag{I4.5}$$

I'll leave it to you to verify, with L'Hospital's limit rule, that for $q/p = p/q = 1$ (that is, for $p = q = 1/2$), the ruin probabilities reduce to

$$P_A = \frac{b}{a+b}, \frac{p}{q} = 1 \tag{I4.6}$$

and

$$P_B = \frac{a}{a+b}, \frac{p}{q} = 1. \tag{I.4.7}$$

You should confirm from (I4.2), (I4.4), and (I4.5) that $\lim_{p \to 1} m = b$ and $\lim_{p \to 0} m = a$, which makes sense because if $p = 1$, then it takes A exactly b games to ruin B, and if $p = 0$, it takes B exactly a games to ruin A.

I.5 MONTE CARLO SIMULATION OF GAMBLER'S RUIN

De Moivre's analysis of gambler's ruin is devilishly clever. But what if you are not a de Moivre, and even if given several years (or even decades!) to think about gambler's ruin, you know deep in your heart that you would never come up with his variable chip-valuation idea? What then? Are you doomed to never know the ruin probabilities, or anything about the duration of play until ruin occurs? The answers were probably yes and yes in de Moivre's day, but not today. Today we have computers, random number generators, and sophisticated number-crunching software like MATLAB®.

In the following code, called **gr.m**, we have a surprisingly brief code that, after being given the values of a, b, and p, simulates the play of gambler's ruin for 100,000 times, all the while keeping track of how long each simulation lasts and which player is ruined in each simulation.

```
gr.m
a=input('What is a?');
b=input('What is b?');
p=input('What is p?');
LS=0;AR=0;
for loop=1:100000
   A=a;B=b;X=0;
   while A&&B>0
      if rand<p
         A=A+1;B=B-1;
      else
         A=A-1;B=B+1;
      end
      X=X+1;
   end
   LS=LS+X;
   if A==0
      AR=AR+1;
   end
end
LS/100000
AR/100000
```

The workings of **gr.m** are pretty straightforward. After being informed of the values of a, b, and p, the variables LS (for "length of game sequence") and AR (for "A is ruined") are each initialized to zero. When **gr.m** is finished, LS will be the total number of individual games played in 100,000 simulations of gambler's ruin, and AR will be the total number of times it was A who was ruined during those simulations. At the start of each gambler's ruin simulation the variables A and B are initialized to a and b, respectively, and the variable X (the current number of games played in the current gambler's ruin simulation) is set equal to zero. An individual gambler's ruin is played in the *while* loop as long as both A and B remain greater than zero. At the end of each game X is increased by one. When the *while* loop is terminated (because either A or B reached zero), LS is increased by X, and AR is increased by one if it was A reaching zero that triggered the *while* loop termination. Then another gambler's ruin simulation is performed. The last two commands in **gr.m** are obvious, giving the average number of games played per gambler's ruin simulation and A's probability of being ruined.

Table I.5.1 shows some simulation results for P_A and m (the average number of games played until A or B is ruined) for some various values of a, b, and p.

In Table I.5.2 the theoretical values for m (equal to ab if $p = q$) and P_A are shown for the same values of a, b, and p, as calculated from (I4.1), (I4.2), (I4.4), and (I4.6).

The agreement between theory and the simulation code is remarkable, but don't think that Monte Carlo simulation is infallible. Suppose,

Table I.5.1. Simulations results

a	b	p	m	P_A
9	1	0.50	9.01	0.1005
90	10	0.50	895.5	0.1013
9	1	0.45	10.99	0.2091
90	10	0.45	764.6	0.8642
99	1	0.45	171.2	0.1817
99	10	0.40	440.6	0.9823
99	1	0.40	160.98	0.3323

Table I.5.2. Theoretical values

a	b	p	m	P_A
9	1	0.50	9	0.1
90	10	0.50	900	0.1
9	1	0.45	11	0.2101
90	10	0.45	765.6	0.8656
99	1	0.45	171.8	0.1818
99	10	0.40	441.3	0.9827
99	1	0.40	161.7	0.3333

for example, that B is infinitely rich, which means that it is simply impossible for A to ruin B. On the other hand, B could certainly ruin A *if p < q*. If *p > q*, however, it isn't hard to show that A is not necessarily ruined even by the infinitely rich B. So, to simulate gambler's ruin with B having arbitrarily huge wealth (with *p > q*) *may*, with non-zero probability, result in a simulation that never terminates. This is a very practical concern!

1.6 NEWTON'S PROBABILITY PROBLEM

In the last section I mentioned that de Moivre was a friend of Newton's. A natural question to ask now is, what did *Newton* think of probability? Newton was one of the world's greatest intellects, with his contributions to science and mathematics everywhere, too numerous to list other than in a full-length biography. And yet, with just two exceptions, he ignored probability. The first exception was a collection of minor contributions to actuarial science (the theory of insurance annuities). The second came only at the prompting of another.

That prompting arrived in the form of three letters, the first dated November 22, 1693, from Samuel Pepys (1633–1703).[10] Pepys is remembered today as the author of a posthumously published *Diary* that gives much insight into life in England in the 1660s. When his letter arrived, Newton knew nothing of Pepys's still private *Diary*, but nevertheless he knew who Pepys was; during the years 1684–1686 Pepys was president of the Royal Society of London, just when Newton's masterpiece, *Principia*, was being prepared for publication in 1687 by the Royal Society. Indeed, Pepys's name as president of the Royal Society, appears on the title page of *Principia*.

Just why Newton seemingly avoided probability theory is still not entirely clear to historians, but one possibility is that the only major applications (in Newton's day) other than in insurance seemed to be those of interest to gamblers. That is, the pursuit of a "calculus of probabilities" would not be one of seeking after pure truth but rather one of mere grubby speculation. Many devout people thought, then, and think now, of gambling as both a failure of good judgment and a diversion of time from more worthy activities. Even worse, gambling is thought by such folks to be downright immoral since it explicitly substitutes for reasoning ability (given to humans by God) a submission to random chance. To gamble is to hope to get something for nothing,[11] an attitude shared with lowlife thieves!

Pepys's own *Diary* (in an entry dated January 1, 1668) gives us a dramatic description of a common gambling scene of the time. There we read of men "cursing and swearing" and "muttering and grumbling," of "gentlemen drunk," and of one man "being to throw seven if he could and failing to do it after a great many throws, [crying] he would be damned if ever he flung seven more while he lived." It is simply impossible to imagine the pious Newton swearing, muttering, and cursing at such a coarse gathering of dice-tossing drunks.[12] So in his initial letter to Newton, Pepys was quite careful to pose his question in a way acceptable to a man of faith, as "an application after truth".

Newton at first found Pepys's original wording of his question to be, as Newton put it, "ill stated." Eventually, however, the two men agreed to the following formulation: of the three events A, B, and C, defined as

A = {a fair die is tossed six times and at least one 6 appears},
B = {a fair die is tossed twelve times and at least two 6s appear},
C = {a fair die is tossed eighteen times and at least three 6s appear},

which of the three events has the greatest chance of occurring?

Newton correctly calculated the chances for each event, and indeed, Galileo fifty years earlier could have solved this problem too, using just the elementary mathematics available to him (simple enumeration). What makes this problem most interesting to us is that while Newton's mathematics is impeccable (no surprise there), at one point his *logical* reasoning is flawed. Oddest of all, this misstep by Newton remained unremarked upon by mathematicians and historians alike until 2006![13]

First, let me go through the way a modern analyst would answer Pepys's question; in the process we'll derive some equations that will let us easily see the flaw in Newton's response to Pepys. If we write p for the probability a die (fair or otherwise) shows a 6 on a single toss, then $1 - p$ is the probability the die doesn't show a 6, and so P_A, the probability of event A, is 1 minus the probability no 6s appear in six tosses:

$$P_A = 1 - (1 - p)^6.$$

In the same way, , the probability of event B, is 1 minus the probability *no* 6's appear in twelve tosses, minus the probability *exactly one* 6 appears in twelve tosses:

$$P_B = 1 - (1 - p)^{12} - \binom{12}{1} p(1-p)^{11},$$

or, with a little arithmetic,

$$P_B = 1 - (1 + 11p)(1 - p)^{11}.$$

And finally, P_C, the probability of event C, is 1 minus the probability no 6s appear in eighteen tosses, minus the probability exactly one 6 appears in eighteen tosses, minus the probability exactly two 6s appear in eighteen tosses:

$$P_C = 1 - (1-p)^{18} - \binom{18}{1} p(1-p)^{17} - \binom{18}{2} p^2 (1-p)^{16},$$

or, with some arithmetic,

$$P_C = 1 - (1 + 16p + 136p^2)(1 - p)^{16}.$$

If we put $p = 1/6$ into these expressions we get $P_A = 0.6651$, $P_B = 0.6187$, and $P_C = 0.5973$. So, *for a fair die, $P_A > P_B > P_C$*, and this is just what Newton told Pepys. This was valuable information for Pepys because he had initially believed that event C was the most likely, and in a letter (dated February 14, 1694) to another of his mathematically minded acquaintances, Pepys revealed that his original motivation for asking his question was, in fact, a gambling concern. There he wrote that it was good he now had the actual chances for the three events as he was "upon the brink of a wager [for 10 pounds, not a trivial amount of money in 1694] upon my former belief." Pepys did have the correct chances now, but it was clear that he still didn't understand what Newton had told him because, in that same letter, he goes on to write that he is afraid of getting involved with future (different) bets, and makes the admission to his friend that Newton's calculations were "beyond my depth."

Perhaps appreciating Pepys's lack of mathematical ability, Newton had at one point attempted to forget the detailed math and to formulate a simple argument that would let Pepys see the result $P_A > P_B > P_C$ in a flash. And it is with that argument that we see Newton stumble. In his first letter (dated November 26, 1693) in reply to Pepys's original query, Newton wrote,

> it appears by an easy computation that the expectation of A is greater than that of B or C, that is, the task of A is the easiest. And the reason is because A has all the chances of sixes on his [dice] for his expectation, but B and C have not all the chances on theirs. For when B throws a single six or C but one or two sixes, they miss of their expectations.

In other words, event A occurs if even a lone six appears on any toss, while both event B and event C require multiple appearances of a six.

Newton's argument makes no mention of the die being fair, and so, if true, it must be true even for a loaded die. But that isn't true! Specifically, let's assume that the die is loaded to make the 6 more likely—say, $p = 1/4$. I'll leave it for you to confirm, but if you plug this value for p into our three earlier equations you'll find that $P_A = 0.822$, $P_B = 0.8416$, and $P_C = 0.8647$. So now $P_A < P_B < P_C$, the reverse of the conclusion for a fair die. Newton's misstep was in considering only the number of possibilities, and not their probability.

When, years later, de Moivre wrote the preface to the first edition of his *Doctrine of Chances* (dedicated, you'll recall from note 9, to Newton), perhaps he had this affair in mind when he said,

> Some of the Problems about Chance having a great appearance of Simplicity, the Mind is easily drawn into a belief, that their Solution May be attained by the meer Strength of natural good Sense; which generally proving otherwise and the Mistakes occasioned thereby being not unfrequent, 'tis presumed that a Book of this Kind, which teaches to distinguish Truth from what seems so nearly to resemble it, will be looked upon as a help to good Reasoning.

I.7 A DICE PROBLEM THAT WOULD
HAVE SURPRISED NEWTON

At the risk of sounding trite, let me start this section by observing that Newton was a pretty smart fellow. There were probably not many dice puzzles around in his day that he wouldn't have been able (eventually) to crack. In 1959, however, a very curious mathematical result was published that is the basis for a disarmingly simple-appearing dice-tossing problem that might have astonished even Newton.

Imagine that we have three independent random variables, X, Y, and Z. If you had asked mathematicians before 1959 whether it would be possible to have $P(X > Y) > 1/2$, $P(Y > Z) > 1/2$, and $P(Z > X) > 1/2$, almost surely a majority would have said no. That's because even mathematicians can easily be seduced by the "naturalness" of transitivity. For example, if we denote the event "football team A outscores football team B" by $A > B$, then $P(X > Y) > 1/2$ means football team X beats football team Y most of the time, and similarly for the other two probability inequalities. Most people, including mathematicians, intuitively think that if X beats Y and Y beats Z, then surely X beats Z. In fact, that's not so: it *is* possible to have three such inequalities. This result is so surprising it is called the *Steinhaus and Trybuła paradox*, after the two Polish mathematicians who discovered it in 1959, Hugo Steinhaus (1887–1972) and Stanislaw Trybuła (1932–2008).

Here's how the paradox appears in the form of what are called *non-transitive dice*. Imagine that we have three identical, perfect cubes. Instead of the usual dots one finds on the faces of ordinary dice, however, the faces of our cubes are inscribed with the numbers 1 through 18, each number appearing on just one face. Let's call our three cubes A, B, and C, and imagine that the numbers are distributed as follows:

A: 18, 9, 8, 7, 6, 5
B: 17, 16, 15, 4, 3, 2
C: 14, 13, 12, 11, 10, 1

By examining each of the 36 ways that A and B can fall, we see that A shows a larger number than does B in 21 of those ways: 18 beats all six of B's faces, and 9 beats three of B's faces, as do 8, 7, 6, and 5. If we assume A and B are fair, with each pair of faces having probability

1/36, then $P(A > B) = 21/36 > 1/2$. In the same way, we can verify that $P(B > C) = 21/36 > 1/2$ and $P(C > A) = 25/36 > 1/2$.

It's perhaps hard to accept, but there it is, and which are you going to believe, the cold, hard math or your lying intuition? It is the mathematical equivalent of M. C. Escher's famous *Ascending and Descending* (1960) and *Waterfall* (1961) drawings in which the paradox of people and water endlessly traveling around and around in a three-dimensional loop is illustrated. Unlike the situation in Escher's drawings, however, our three dice actually exist.

One can easily imagine how Pepys would have drooled, dreaming of all the suckers he could have relieved of their money if only he had known of nontransitive dice! Newton, on the other hand, with his piety, would no doubt have been simply appalled at how God could have let such an outrageous thing happen.

I.8   A COIN-FLIPPING PROBLEM

As I mentioned earlier, difference equations often appear in probability, and here's one quick example of that (a second example is given in the next section). This problem is too short to warrant a full entry in the book but also too good to leave out. (It would be very interesting to know what Newton would have done with it.) If you flip a coin n times, with probability p of heads on each independent flip, what is the probability of getting an even number of heads? This is an easy question to answer for small n; for $n = 1$ the answer is $1 - p$, since the single flip produces a tail with that probability (and so the number of heads is zero, and that's even), and for $n = 2$ the answer is $(1 - p)^2 + p^2$, since the double flip produces two tails (zero heads) or two heads with that probability. For larger n, however, enumeration quickly becomes overwhelming. We need a better approach.

Let $P(n)$ be defined as the probability of the event "even number of heads in n flips." Then $1 - P(n)$ is the probability of the complementary event, "odd number of heads in n flips." Now, every sequence of heads (H) and tails (T) of length n symbols starts either with an H or a T. If it starts with a T (with probability $1 - p$), then we have $n - 1$ symbols left with which to get an even number of heads. If it starts with an H (with probability p), then we have $n-1$ symbols left with which to get an odd number of heads. So,

$$P(n) = (1-p)\,P(n-1) + p[1-P(n-1)].$$

That is, we have the first-order, linear, *inhomogeneous* difference equation

$$P(n) = p + (1-2p)\,P(n-1),\ P(1) = 1-p.$$

Notice that this says

$$P(1) = p + (1-2p)\,P(0) = 1-p.$$

And so $P(0) = 1$, which makes sense. That is, if we haven't flipped the coin at all, we are *absolutely guaranteed* to have zero heads—which is even!

So, how do we solve our difference equation? In general, with a constant term (the p) on the right-hand side, the general solution will have the form of a constant plus a power term. So, let's assume

$$P(n) = k + Ca^n$$

where k, C, and a are all constants. Substituting this assumed solution into the difference equation, we have

$$k + Ca^n = p + (1-2p)\,[k + Ca^{n-1}].$$

Equating the constants on each side of this equation, and equating the power terms on each side of this equation, we have

$$k = p + (1-2p)\,k$$

and

$$Ca^n = (1-2p)\,Ca^{n-1}.$$

These are easily solved to give $k = 1/2$ and $a = 1-2p$. Thus,

$$P(n) = \frac{1}{2} + C(1-2p)^n.$$

Finally, using $P(1) = 1 - p$, we find that $C = 1/2$. So, the answer to our question is that

$$P(n) = \frac{1}{2} + \frac{1}{2}(1-2p)^n.$$

For a fair coin the power term vanishes and $P(n) = 1/2$ for any n, while for a biased coin $P(n)$ generally varies with n as long as $p \neq 0$ (if $p = 0$, then $P(n) = 1$, *always*, since in that case you always get tails and so zero heads). If $p = 1$ (the coin always shows heads) then $P(n)$ is 0 for odd n and 1 for even n. And for any p other than 0 or 1, $\lim_{n \to \infty} P(n) = 1/2$.

This particular coin-flipping problem is a random process that is easy to simulate, as does the code **flip.m**. The operation of the code is straightforward. Once the values of p and n are set, the variable *even* is initialized to zero (when the code terminates, *even* will equal the number of times, out of one million simulations of n flips, that resulted in an even number of heads).

---

flip.m
```
p=0.1;n=9;even=0;
for loop1=1:1000000
    for loop2=1:n
        result=rand;
        if result<p
            f(loop2)=1;
        else
            f(loop2)=0;
        end
    end
    total=sum(f);
    test=2*floor(total/2);
```

(continued)

(continued)

```
        if total==test
            even=even+1;
        end
    end
    even/1000000
```

A simulation of *n* flips of the coin is done by generating *n* numbers, uniformly random from 0 to 1, and storing a 1 in vector *f* for each number that is less than *p* and a 0 for each number that is greater than *p*. A summation of the elements of *f* will give the number of heads in the *n* flips. This value (the variable *total*) is tested for "evenness" by simply dividing it by two and rounding the result down (with the command *floor*: *floor*(3.3) = 3, while *floor*(3) = 3), and then multiplying that by two to get *test*. Only if the number of heads is even will *total* equal *test*. If that's the case, then *even* is incremented by one, and then another *n* flip simulation is performed. When all one million simulations are done, the last line of **flip.m** prints the code's estimate for the probability of an even number of heads. If $n = 9$ and $p = 0.1$, the theoretical probability is 0.5671, while the code's result was 0.5669. If $n = 9$ and $p = 0.9$, the theoretical probability is 0.4329, while the code's result was 0.4326. Pretty good agreement.

In one of the puzzles later in this book I'll remind you of what we did here, and I think you'll find it quite helpful.

I.9   SIMPSON'S PARADOX, RADIO-DIRECTION FINDING, AND THE SPAGHETTI PROBLEM

To end this introduction, I'll leave "old history" and tell you three "new history" stories. The first one has to do with statistics, a very close relative of probability. Statistical issues really make no appearance in this book, which some may view as a disappointment, so perhaps this section will earn me some partial forgiveness. Imagine two hypothetical baseball players, A and B, who have just completed their first two years as professionals. Their at-bats and hitting numbers for each of those two years and their two-year combined totals are as follows (with their batting averages in bold):

Player	Last year		This year		Combined	
A	13/52	(.250)	179/581	(.308)	192/633	(.303)
B	108/409	(.264)	46/138	(.333)	154/547	(.282)

Notice that B has the higher batting average in each of the two years, but nonetheless the lower average when the two years are combined!

This sort of behavior is generally thought by most people to be quite strange; even mathematicians, who are used to seeing some pretty pathological behaviors, have tagged it as a paradox. Specifically, it is an example of *Simpson's paradox*, named after the British statistician Edward Hugh Simpson (born 1922), who wrote of it in a 1951 paper (even though the Scottish statistician George Udny Yule (1871–1951) had much earlier commented on it, in 1903). Simpson's paradox typically occurs when data sets of unequal size are combined into one large set, as is often done in medical studies. The results often are conflicting conclusions concerning the value of various medical treatments being studied, which typically leads to confusion rather than edification. My next example of Simpson's paradox shows, quite dramatically, how that can happen.

Suppose a new, previously unknown infection has appeared for which two new drugs are being tested at two different hospitals. Hospital 1 gives the drugs to a total of 230 men, with the success/failure outcome as follows:

Drug	Success	Failure
1	60	20
2	100	50

With these results, drug 1 has a success rate of $60/80 = 0.75$, while drug 2 has a success rate of $100/150 = 0.67$.

Hospital 2 gives the drugs to a total of 160 women, with the outcomes as follows:

Drug	Success	Failure
1	40	80
2	10	30

With these results, drug 1 has a success rate of $40/120 = 0.33$, while drug 2 has a success rate of $10/40 = 0.25$.

Both hospitals therefore reach the same conclusion: drug 1 is the better drug. But is it?

Suppose we combine the two tests, to have a total of 390 people. With drug 1 we have 100 successes and 100 failures (a success rate of $100/200 = 0.5$), and with drug 2 we have 110 successes and 80 failures (a success rate of $110/190 = 0.579$). Now drug 2 appears to be the better drug. What should a person conclude? Take two aspirin and go to bed!

For my second story, I'll start by reminding you of a scene we've all seen at least once in some action/drama movie of World War II. The scene opens in a tiny room barely lit by the flame of a flickering candle—perhaps it's a basement closet or an attic—with a man hunched over a radio transmitter Morse code key. He is rapidly sending an encrypted message from the underground resistance in (pick a city, pick a country) to incoming British SAS commandos concerning the sudden shift in the whereabouts of some important Nazi official who has been flagged for "termination with extreme prejudice."

Suddenly the scene shifts to the local Gestapo headquarters, and we hear the excited yelling of a man wearing headphones who has just picked up the underground's radio signal. His boss, a brutal-looking SS colonel wearing a suit and black gloves, strides over to a detailed street map of the city that is hanging on a wall beneath portraits of Adolf Hitler and Heinrich Himmler. "Sergeant," he barks at the radio man, "tell the hunter-vans to give me fixes. Quickly!"

The scene shifts again, to a speeding, black Gestapo radio intercept van, and we see it suddenly pull over and stop, and then a circular antenna on the roof begins to slowly rotate. There is a close-up shot of the antenna's angular-bearing dial inside the van. We listen in as the sergeant back at Gestapo headquarters receives the report, "Intercept bearing from (pick a street) of 72 degrees." He tells the colonel. The colonel, still in front of his map, grunts an acknowledgment, then pins one end of a thin red thread to the map at the van's location and stretches it tight across the map at an angle of 72 degrees. Then a second radio intercept van reports in with a new angle from a different street location, and a second thread is similarly pinned to the map. The colonel taps the map where the threads intersect, chuckles evilly, and says, slowly and ominously, "*Now* we have you."

The final scene is of a troop carrier roaring up to a building, discharging two dozen soldiers, each armed with a nasty-looking MP-40 machine pistol. Things look pretty bleak for the hero.

Would this little drama actually play out as I've described? Sort of, but not exactly. In Figure I.9.1 the point *P* is the actual location of the underground's transmitter. When the first intercept van reports in, small but unavoidable errors in both the van's location and the measured bearing angle will result in the colonel's first thread (shown in the figure as line *A*) almost certainly not passing through *P* but rather having some small displacement. When the second van reports, the same situation results, giving line *B* also not passing through *P*, and so the intersection point of *A* and *B* is almost certainly not *P*. What might be done is to have a third van report, giving line C in the figure, a line similarly displaced from *P*. The Gestapo therefore hasn't located *P* exactly, but it does seem to have at least put *P* somewhere inside the small triangle *abc*. That would mean soldiers could certainly cordon off that entire area and then begin a detailed search inside the triangle.

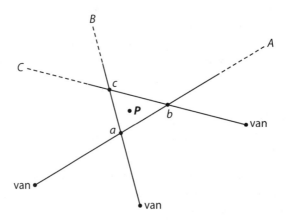

Figure I.9.1. The geometry of radio-direction finding.

But is *P* really inside the triangle? Maybe, but it's actually more likely that *P* isn't there! This was shown in a 1946 paper by the physicist Samuel Goudsmit (1902–1978), who presented a very simple probability argument to show the surprising result there is "a three-to-one chance that the true position is outside" the triangle.[14] Here's how Goudsmit reasoned.

As shown in the figure, unavoidable errors have caused *A* to fall to the right of *P*. There is an equal probability that *B* falls on the right or the left of *P* (I've drawn it to the left). The position of *C* shows that *P* can be inside the triangle only if *B* passes *P* on the left; the probability of that is 1/2. And finally, when *C* is drawn you can see that *P* is inside the triangle only if *C* passes above *P*. That has, again, probability 1/2. Thus, the probability that *P* is inside the triangle is 1/4, and so the probability is 3/4 that *P* is actually outside the triangle, three times the probability that it is inside. Maybe our underground radio operator survives after all!

My final tale is of a recently solved probability question with a most surprising answer. Once again, a difference equation will be the key to our solution. Imagine *n* strands of spaghetti in a boiling pot of water. At all times, all you can see are the free ends of the spaghetti strands sticking up through the water's surface. You choose two of the visible ends at random, tie them together (this is a pure math problem, so realism isn't a top-priority issue here), and drop the connection back into the water (where it is not visible). You keep doing this until no more free ends are visible. You then pour the pot of water through a strainer. How many spaghetti loops do you expect to find in the strainer? If we call the answer $E(n)$, then obviously $E(1) = 1$. If you do a little exhaustive enumeration, you should be able to convince yourself that $E(2) = 4/3$. For $n > 2$, however, enumeration becomes very difficult, very quickly (just for fun, try your hand at the $n = 3$ case). So, let's derive a theoretical expression for $E(n)$ and evaluate it for $n = 100$, for $n = 1,000$, and for $n = 10,000$. It should be physically obvious that $\lim_{n \to \infty} E(n) = \infty$, but I think you'll be surprised by how small the results are for even large *n*.

This problem and its solution have an interesting origin. In December 2009 I flew out to Monterey, California, to give a talk at a math conference. Just before the talk I came across the spaghetti problem (posed as possibly an unresolved problem) in Steve Strogatz's new book, *The Calculus of Friendship* (Princeton 2009), and I made passing mention of it during my presentation. Two days after I returned home, I received the following elegantly simple solution in an e-mail from Matt Davis (on the math faculty at Chabot College), who had been in the audience.[15]

Starting with n strands of spaghetti, there are $2n$ ends sticking up through the water's surface. Randomly choosing one of those ends, we then have $2n-1$ ends remaining for the second choice. There are now two possibilities. First, with probability $\frac{1}{2n-1}$ the second choice is the other end of the same strand as the first choice. We then have a completed loop, and so $n-1$ strands are left in the pot. With this outcome we have the expected number of loops as $1 + E(n-1)$. The second possibility, with probability $\frac{2n-2}{2n-1}$, is that the second choice is an end from a strand different from the first strand. After we tie this second choice to the first choice we simply have $n-1$ pieces of spaghetti in the pot (one is now longer than all the others, but that's irrelevant—all we see are the ends). So, with this outcome we have the expected number of loops as $E(n-1)$.

Weighting each of these two expectations of the number of loops by their probabilities to get the overall expected number of loops, we (remembering that $E(1) = 1$) arrive at the nonlinear difference equation

$$E(n) = \frac{1}{2n-1}[1 + E(n-1)] + \frac{2n-2}{2n-1}E(n-1)$$
$$= \frac{1}{2n-1} + \left[\frac{1}{2n-1} + \frac{2n-2}{2n-1}\right]E(n-1),$$

or

$$E(n) = E(n-1) + \frac{1}{2n-1}.$$

Even though nonlinear, this is actually easier to solve than was the linear difference equation in the previous section's coin-flipping problem. So,

$$E(2) = E(1) + \left(\frac{1}{3}\right) = 1 + \left(\frac{1}{3}\right),$$
$$E(3) = E(2) + \left(\frac{1}{5}\right) = 1 + \left(\frac{1}{3}\right) + \left(\frac{1}{5}\right),$$
$$E(4) = E(3) + \left(\frac{1}{7}\right) = 1 + \left(\frac{1}{3}\right) + \left(\frac{1}{5}\right) + \left(\frac{1}{7}\right),$$

and so on. In general, $E(n) = \sum_{k=1}^{n} 1/(2k-1)$. Thus, the partial sums of $E(n)$ grow even more slowly than do those of the harmonic series, which itself has very slowly increasing partial sums. Specifically, $E(100) = 3.28$, $E(1,000) = 4.44$, and $E(10,000) = 5.59$, numbers that I personally find counterintuitive in their "smallness." Because of that, in fact, the spaghetti problem is a good example of a well-defined random physical process that is *not* a good candidate for study by computer simulation. A change in $E(n)$, even for a "big" change in n, will be swamped by statistical sampling fluctuations.

Nearly a hundred years ago an out-of-sorts physicist grumbled in a paper that "Mathematicians of today are like moles, each in his little burrow finding his way about apparently by a sixth sense and turning up [his own] rubbish heap." It's easy to understand that frustration, as so much of modern math *is* understandable only to specialists who alone can appreciate what their subject is even about. There will be no doubt in your mind about what the following probability problems are about, however, and I think even that long-ago physicist would have had fun with what is in this book. Enjoy!

NOTES AND REFERENCES

1. The relationship between Gombaud and Pascal is often misstated in probability textbooks, as is carefully and entertainingly discussed in Oystein Ore, "Pascal and the Invention of Probability Theory" (*American Mathematical Monthly*, May 1960, pp. 409–419). Pascal is famous in the philosophical community for the so-called *Pascal's wager*, which appeared in his posthumously published (1669) *Pensées* (*Thoughts*). This was his use of mathematical expectation to conclude it was in the interest of even otherwise confirmed atheists to proclaim a belief in God. This argument does have the glaring flaw of ignoring the fact that an omniscient being would certainly know of the spiritual emptiness of the proclamation! You can find more on Cardano in Ore's book, *Cardano, the Gambling Scholar* (Princeton, N.J.: Princeton University Press, 1953).

2. Oscar Sheynin, "Stochastic Thinking in the Bible and the Talmud" (*Annals of Science*, April 1998, pp. 185–198). The Bible contains many references to lotteries in both the Old and the New Testaments. For example, when a choice had to be made between Barsabas and Matthias as the successor to the apostle Judas Iscariot, it wasn't done by direction from God but instead by lot (Acts 1:23–26). An even more dramatic example is found in the four gospels, which are not infrequently in conflict. On one point, however, they do agree: when

the Roman soldiers present at the Crucifixion divided the garments of Jesus, they did so by lot (Matthew 27:35, Mark 15:24, Luke 23:24, and John 19:23–24).

3. This assumption is a circular one, as it implicitly assumes that one already knows what a probability is even as we use it to define probability! Nonetheless, it is in that famous category of "you know what I mean," and so the circular objection is usually taken seriously only by the purest of the pure.

4. A.W.F. Edwards, "Pascal and the Problem of Points" (*International Statistical Review*, December 1982, pp. 259–266).

5. A.W.F. Edwards, "Pascal's Problem: The 'Gambler's Ruin'" (*International Statistical Review*, April 1983, pp. 73–79). This problem is equivalent to another with a less provocative, more serious name: the "random walk with two absorbing barriers."

6. Paul J. Nahin, *Mrs. Perkins's Electric Quilt and Other Intriguing Stories of Mathematical Physics* (Princeton, N.J.: Princeton University Press, 2009, pp. 241–245). My difference equation discussion is for a special case; for a more general difference equation analysis see volume 1 of William Feller, *An Introduction to Probability Theory and Its Applications*, 3rd ed. (New York: John Wiley, 1968, pp. 344–349).

7. Eddie Shoesmith, "Huygens's Solution to the Gambler's Ruin Problem" (*Historia Mathematica*, May 1986, pp. 157–164).

8. A. R. Thatcher, "A Note on the Early Solutions of the Problem of Duration of Play" (*Biometrika*, December 1957, pp. 515–518).

9. The usual textbook derivation of the ruin probabilities is the one that appears in the famous, posthumously published 1713 book, *Ars conjectandi* (*The Art of Conjecturing*), by James (aka Jacob) Bernoulli (1654–1705), using the difference equation approach (see note 6 again). De Moivre's derivation was actually in print first, however, appearing in the essay "De Mensura Sortis" ("On the Measurement of Chance") in a 1711 issue of the *Philosophical Transactions of the Royal Society of London*, and then again in de Moivre's 1718 book, *Doctrine of Chances*. (The first edition of *Doctrine* is dedicated to Newton.) De Moivre's work gets a nice tutorial review in the paper by the Dutch mathematician Anders Hald (1913–2007), "A. de Moivre: 'De Mensura Sortis or 'On the Measurement of Chance,'" *International Statistical Review*, December 1984, pp. 229–262.

10. You can find these letters reprinted in Florence N. David, "Mr. Newton, Mr. Pepys & DYSE: A Historical Note" (*Annals of Science*, September 1957, pp. 137–147), and Emil D. Schell, "Samuel Pepys, Isaac Newton, and Probability" (*American Statistician*, October 1960, pp. 27–30). If that "DYSE" in David's *Annals of Science* title is puzzling you, that's part of the reason why it's not always easy to read letters that are hundreds of years old—that's the old-fashioned spelling for "dice." David (1909–1993), late professor of statistics at the University of California, Berkeley, wrote a very nice book-length history (up to the time of Newton) of probability in 1962 that is well worth the effort to locate today (*Games, Gods and Chance*).

11. There is a vivid example of what I mean by this in Lorraine Daston, *Classical Probability in the Enlightenment* (Princeton, N.J.: Princeton University Press, 1988, p. 160), which describes the "motley following" of lotteries: "one stream of *Coachmen, Footmen, Prentice Boys*, and *Servant Wenches* flowing one way, with wonderful hopes of getting an estate for three pence. *Knights, Esquires, Gentlemen* and *Traders, Marry'd Ladies, Virgin Madams, Jilts*, etc.; moving on *Foot*, in *Sedans, Chariots*, and *Coaches* another way; with a pleasing Expectation of getting Six Hundred a Year for a Crown." The original letter Pepys wrote to Newton was actually prompted by Pepys's involvement with such a speculator, John Smith, who was writing master at Christ's Hospital in 1693. It was Smith who originally formulated the question, and when he asked Pepys (who was a governor of the hospital) about it, Pepys in turn asked Newton. You can find more on this in the paper "John Smith's Problem," by T. W. Chaundy and J. E. Bullard (*Mathematical Gazette*, December 1960, pp. 253–260).

12. Gambling in Newton's day did have the interesting feature of bringing together, at least for a while, all manner of loose, shady characters from vastly different walks of life and classes of society. As Pepys reported in his *Diary* (in an entry dated December 21, 1663), while watching a cockfight, "Lord! To see the strange variety of people, from [a] Parliament man . . . to the poorest 'prentices, bakers, brewers, butchers, draymen, and what not; and all these fellows one with another cursing and betting."

13. Stephen M. Stigler, "Isaac Newton as a Probabilist" (*Statistical Science*, August 2006, pp. 400–403).

14. S. A. Goudsmit, "Accuracy of Position Finding Using Three or Four Lines of Position" (*Navigation*, June 1946, pp. 34–35). An editorial introduction to this paper told readers that Goudsmit was in charge of the theoretical group at the MIT Radiation Laboratory (working on radar) "during the early part of the war, until a confidential mission took him to the European theater." Nothing more was said, but it was later revealed that Goudsmit's mission was as scientific head of Alsos, the cover name for a group of analysts who immediately followed the Normandy invasion forces after the D-Day landings on the coast of France in an attempt to discover the details of the German atom bomb project. (This seemingly odd name was actually an inside joke: alsos is the Greek word for groves, in honor of General Leslie Groves, who was head of the *American* atom bomb project!)

15. The original statement of the spaghetti problem (with strings in a box instead of spaghetti strands in a pot), and its solution, was made by two Australian electrical engineers: see their paper, "How Many Loops in the Box?" (*Mathematical Gazette*, March 1998, pp. 115–118), by Dushy Tissainayagam and Brendan Owen.

Challenge Problems

Before starting with the formal problems in this book, here are several challenge problems for you to think about as you read. As you work your way through the formal problems you may pick up ideas and mathematical techniques that will serve you well in attacking the challenge problems. These are not necessarily easy problems (you'll have to do some integrals), and for most of them, you will have to think long and hard. All do have exact solutions, and you will find complete discussions (including Monte Carlo simulations) of each at the back of the book. But I strongly encourage you to work hard at these problems before you look there.

(1) Imagine all possible triangles in which two of the sides are independently and randomly (uniformly) assigned some length from 0 to 1. From all these triangles, consider only the set S in which the third side has the exact length of 1. If we select a triangle at random from S, what's the probability it is obtuse?

(2) In the 1980s, variations of the following puzzle, called the glass rod problem, made several appearances in the British math journal *The Mathematical Gazette*. It was originally posed and solved in the March 1981 issue as follows.

A glass rod drops and breaks into three pieces. What is the probability that a triangle can be formed from the pieces?

Consider an equilateral triangle whose height is the length of the rod (say 1). It is easily proved that the sum of the lengths of the perpendiculars from the three sides to any interior point is 1 [no proof was offered, but this is often called *Viviani's theorem*, after the Italian mathematician Vincenzo Viviani (1622–1703), to whom it is attributed; it is not in Euclid's *Elements*]. Further, given any triple of non-negative numbers whose sum is 1, there is just one interior point

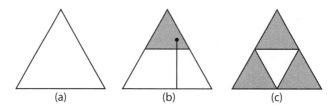

Figure C.1. The glass rod problem.

whose distances from the three sides (in some prescribed order) are precisely those three numbers [these distances are the lengths of the three pieces]. Hence there is a 1:1 correspondence between the interior points of the triangle and the possible lengths of the three pieces of the broken rod (see (*a*) in the figure). If the perpendicular from the base extends into the shaded region in (*b*), then its length is over 1/2, and so the point shown corresponds to a failure to form a triangle [to form a triangle, the lengths of all three pieces must, of course, be less than 1/2]. But the same applies to the other two perpendiculars and so there are three failure regions, each 1/4 of the triangle (as shown in (*c*)). So the probability of being able to form a triangle with the three pieces is 1/4 (= 1 − 3/4).

Well, that's certainly slick! But are you convinced? By taking the ratio of the unshaded area to the area of the entire triangle, the central assumption of geometric probability is being invoked: each tiny patch of area in the triangle is as likely as any other tiny patch (of the same area) to be selected as the location of an interior point. Is that the case here? Now, don't read too much into my question: the above analysis may be faulty or it may be okay—I'm simply asking you to think carefully about both the statement of the problem and the possible unstated assumptions being made in the analysis. This is important, because it will be quite useful in a later challenge problem.

(3) After you've thought about the previous problem for a while, here's what seems to be the same problem, except that the details of the breaking of the glass rod are (much) more

carefully specified. The rod is of unit length, with the left end at 0 and the right end at 1. We break it as follows: we independently and randomly (uniformly) pick two breakpoints at distances x and y, in the interval 0 to 1. This gives us two possibilities for the length of the broken pieces, one possibility for $x < y$ and another for $x > y$. If $x < y$, then the three pieces have lengths x, $y - x$, and $1 - y$. If $x > y$, then the three pieces have lengths y, $x - y$, and $1 - x$. What is the probability the three pieces can form a triangle?

(4) As a continuation of the previous problem, suppose a triangle *is* formed. Given that, what's the probability it's an obtuse triangle?

(5) One last broken rod. First break the rod at a random point selected from the interval 0 to 1. Then take the longer of the two resulting pieces (there will always be a short piece and a long piece after the first break) and break it at some point randomly (uniformly) selected along its length. What is the probability the three resulting pieces can form a triangle?

(6) Getting tired of breaking rods and making triangles? Let's toss darts instead, then. Suppose you have a circular dartboard with unit radius, and you toss two darts at it. Both darts hit the board at independent, random locations. As usual, assume the randomness is uniform. What's the probability the two darts are at least unit distance apart?

(7) Given a circle centered on the origin, pick three points on its circumference at random. By random, I mean use the following selection process for the three points. First, pick any point on the circumference, then rotate the circle so as to put that point on the positive x-axis. That sets the first point, and there is no loss in generality in doing this because of the symmetry of the circle. Then, independently pick two angles, α and β, each uniform from 0 to 2π. The second point is located from the first point by angle α (going counterclockwise), and the third point is angle β from the first point (also going counterclockwise). Now, using those points as the vertices of a triangle, what's the probability the center of the circle (the origin) will be inside the triangle? *Hint:* You might

want to simulate this process first, and then do a theoretical analysis.

(8) Six sports teams, all of equal ability, form a league and play a long series of games each year to determine the overall champion. Each year the championship team receives a glorious trophy. If the same team wins the trophy three straight years, that team gets to retire the trophy permanently. How many years do these six teams have to play before the probability is at least 1/2 that the trophy is retired? How does the answer change if the league consists of ten teams? (*Hint:* This problem can be formulated as a difference equation that can be solved analytically, but you may find a computer approach to be vastly easier.)

(9) Imagine that a golfer has hit a long drive that has randomly placed his ball on a perfectly flat, square green, with the hole in the center of the green. What is the probability that the ball is closer to the hole than it is to any edge of the green?

(10) Imagine an urn that initially contains b black balls and w white balls, where b, $w > 0$. You randomly draw a ball, note its color, and then discard it. You then randomly draw more balls, one after the other, and, as long as their color matches that of the first ball, you discard each of them, too. Once you draw a ball of the other color, however, you put it back into the urn. You then repeat this entire process (the color of the next ball drawn is noted, that ball is discarded, . . .). Finally, there is just one ball left in the urn. When you draw it, what's the probability it is black?

(11) This book has already included a fair amount of historical discussion dealing with probability problems involving the tossing of dice, so here's one more for you as the penultimate challenge problem. The Swiss mathematician James (aka Jacob) Bernoulli (1654–1705) formulated the following question in 1685, and discussed it in 1690 on the pages of the early scientific journal *Acta Eruditorum* (Reports of the Scholars). Two players, A and B, take turns tossing a fair die. The first one to throw an ace (the face with a single dot) wins. They proceed as follows. On the first turn, A throws

once, then B throws once if A didn't throw an ace. For the second turn, if B didn't throw an ace, then A gets the die back and throws twice, and then B throws twice if A didn't get an ace. And so on, with each new turn allowing each player one additional toss. That is, on the kth turn with the die, A gets k tosses, and then so does B if A didn't throw an ace. What's the probability that A wins? (Bernoulli found an infinite series for the answer, but didn't evaluate it.)

(12) For the final challenge problem, here's one that is at the level of a final exam question in a first-year college course in probability. It's easy to understand and very easy to simu-late, but to do it analytically you'll really have to understand the fundamentals of probability theory. So, with that big buildup, here it is. Suppose we pick two numbers indepen-dently, with each uniformly distributed over the interval -1 to 1. Call them A and B. What's the probability that $A^{2/3} + B^{2/3} < 1$? To simulate this is duck soup, and here's a MATLAB® code that does the job:

---

final.m
```
s=0;p=1/3;
for loop=1:1000000
    A=(-1+2*rand)^2;B=(-1+2*rand)^2;
    A=A^p;B=B^p;
    if A+B<1
        s=s+1;
    end
end
s/1000000
```

When run several times, the code produced estimates for the proba-bility ranging from 0.2937 to 0.2957. Calculate the exact value of this probability. (By the way, you'll notice that after finding A and B, the code first squares them and then takes the cube root. That is, of course, the two-thirds power, but why not just compute $A^{2/3}$ and $B^{2/3}$ directly?

I did the calculations this way because if A (or B) is less than zero, that number raised to the two-thirds power is positive, and doing the calculations as done in the code is guaranteed to produce a positive result. Writing $A^{2/3}$ and $B^{2/3}$ directly will not work because 2/3 in a computer is 0.66666666 . . . to a *finite* number of digits, with the result being complex results.)

Breaking Sticks

1.1 THE PROBLEM

For our first puzzler, let's start by considering an easy warm-up question. Then, by making just a very slight (or so it will at first seem) alteration, we'll have a new question that will produce what I think will be a surprising result for you. This new question will still not be all that difficult to answer theoretically, but the result will be sufficiently surprising that a computer simulation will be of real help in convincing you that we haven't made a mistake somewhere. That said, here's the warm-up.

Suppose you have a stick that is of unit length (in whatever units you like). Imagine that somebody makes two invisible marks on the stick, each mark made independently and at random. The words "independently and at random" have a quite precise mathematical meaning; specifically, each mark is selected uniformly along the stick from 0 (the left end) to 1 (the right end), and the location of each mark has no connection with where the other mark is made. Next, a second person, with no knowledge of where the two invisible marks are, breaks the stick into n equal length pieces. What's the probability that the two invisible marks are on the same piece?

1.2 THEORETICAL ANALYSIS

The answer is $1/n$, and here's why. Clearly, one of the marks has to be on one of the pieces. The probability that the other mark, as likely to be on one piece as on any other, is on the same piece is (equally obviously)

$1/n$. *That's it*—we are done! Now, here's a little alteration to make things a bit more interesting. The two invisible marks are made just as before, but now the second person doesn't break the stick into n equal length pieces but rather selects $n-1$ locations independently and at random, marks those locations, then breaks the stick at those locations. We again have n pieces, but now they are not necessarily of equal length. What now is the probability that the two invisible marks made by the first person are on the same piece?

Even if you can't see the answer yet, do you have a feeling for whether the answer is greater or less than the first answer, $1/n$? Think about this before reading on, and then see if you're surprised. I was.

The two invisible marks, and the $n-1$ break locations, together make a total of $n+1$ marks, each independently and randomly located along the stick. Call the leftmost mark no. 1, the mark immediately to right of it no. 2, and so on, all the way to the rightmost mark, which will be no. $(n+1)$. There are then a total of $\binom{n+1}{2}$ different ways to select two of the $n+1$ marks to be the two invisible marks, where $\binom{x}{y}$ is the binomial coefficient $x!/(x-y)!y!$, with x and y both non-negative integers and $y \le x$. Now, here's the crucial observation: for these two invisible marks to be on the same piece of the broken stick, they must be adjacent marks (that is, not separated by one or more of the break marks). This means the invisible marks must be either no. 1 and no. 2, or no. 2 and no. 3, or no. 3 and no. 4, or . . . or no. n and no. $(n+1)$. This is a total of n different possible pairings.

So, we immediately have our answer: the probability that the two invisible marks are on the same piece is

$$\frac{n}{\binom{n+1}{2}} = \frac{n}{(n+1)!/2!(n-1)!} = \frac{2n}{(n+1)n} = \frac{2}{n+1}.$$

That is, if we break the stick randomly, then, for large n, we almost double the probability that the two invisible marks are on the same piece compared to when we broke the stick into pieces of equal length. Even if you thought the probability would increase, did you think it would increase by that large a factor?

If you find this problem too easy, then try your hand at this extra-credit question, which is a simple model for how radiation-damaged chromosomes might repair themselves. (Chromosomes are long, threadlike structures in cells that are the carriers of the genes, which are strung out along the length of the structure; genes are the carriers of the information that distinguishes us as individuals, from eye color to nose shape to predispositions to diseases. A broken chromosome, then, is a worrisome thing, representing damage to a cell.) Suppose we break each of n sticks (think of chromosomes) into two pieces and, to give them names, let's call the two parts from each broken stick the long part and the short part. We now have $2n$ parts in a big pile. Then we randomly glue the parts back together, in pairs. For each of the glued pairs there are three possibilities: long + long, long + short, and short + short. What is the probability that every one of the resulting n glued pairs is a long part glued to a short part? (It is not required that any of these glued pairs reunite a long part with its original short part.)

1.3   COMPUTER SIMULATION

To check our theoretical result of $2/(n+1)$ for the second question, let's next do a Monte Carlo simulation called **marks.m**. Here's how it works. After getting the value of n, the number of pieces into which the stick will be broken, the code begins by initializing the variable *same* to zero, which will be increased by one each time a simulation results in the two invisible marks ending up on the same piece. Then the first of one million simulations begins. The row vector x receives the locations of $n-1$ random locations for where the stick will be broken, and then the row vector *cut* sorts the values of x into ascending order, from left to right. That is,

$$0 < cut(1) < cut(2) < cut(3) < \ldots < cut\,(n-1) < cut\,(n) = 1.$$

Note carefully that $cut(n)$ is not an actual break location but simply the right end of the stick. The random locations of the invisible marks are $y(1)$ and $y(2)$.

Then, with the variables *look* and k both initialized to 1, the *while/end loop* determines which piece of the broken stick the invisible mark at $y(1)$ is on, and *remember* is assigned that piece number. The final *if/else/*

end loop then checks to see whether the invisible mark at *y(2)* is on the same piece. When the last simulation is completed, the value of *same* is the total number of times the two invisible marks were on the same piece, and so *same/1000000* is the code's estimate of the probability we are after. Table 1.3.1 compares theory with experiment for $2 \leq n \leq 9$. As you can see, the agreement is pretty good.

---

marks.m

```
n=input('How many pieces (=>2)?')
same=0;
for trials=1:1000000
    for loop=1:n-1
        x(loop)=rand;
    end
    cut=sort(x);cut(n)=1;
    y(1)=rand;y(2)=rand;
    look=1;k=1;
    while look==1
        if y(1)>cut(k)
            k=k+1;
        else
            remember=k;look=0;
        end
    end
    if remember==1
        if y(2)<cut(remember)
            same=same+1;
        end
    else
        if y(2)<cut(remember)&&y(2)>cut(remember-1)
            same=same+1;
        end
    end
end
same/1000000
```

Table 1.3.1. Theory versus Experiment

n	theory	simulation
2	0.666666	0.666789
3	0.5	0.500083
4	0.4	0.399306
5	0.333333	0.334058
6	0.285714	0.285119
7	0.25	0.249765
8	0.222222	0.221585
9	0.2	0.199694

Now, how about the extra-credit problem on the n broken sticks? Imagine that, as we break each stick, we paint a number on each part, starting with 1 on the long part of the first broken stick and 2 on the short part of the first broken stick. The long part of the second broken stick gets a painted 3 and the short part gets a painted 4, and so on. This process will, when we finish breaking all n sticks, give us $2n$ parts, each painted with a unique number. In other words, the $2n$ parts are *distinguishable*. We next start to randomly glue the parts together, in pairs. I'll use the following imagery to describe how we'll do this.

Let there be $2n$ boxes in front of us, lined up from left to right. Into each box we'll put one part. Then, taking the boxes two at a time from left to right, we'll glue the parts in the two boxes together. Define N_1 as the total number of *distinguishable* ways to place $2n$ parts in $2n$ boxes, one part in each box. For example, if the parts painted 19 and 7 end up in a given pair of boxes, the order 19,7 is *distinguishable* from the swapped order 7,19, even though both possibilities represent the same two parts being glued together. Also, define N_2 as the total number of distinguishable ways each pair of boxes gets one long part and one short part. The answer to our question is then N_2/N_1.

To calculate N_1, we start with the leftmost box and put any one of the $2n$ parts in it. The next box to the right gets any one of the remaining $2n - 1$ parts, and so on. So,

$$N_1 = (2n)(2n-1)(2n-2)\ldots(3)(2)(1) = (2n)!$$

To calculate N_2, notice that the first two boxes must receive one long part and one short part and, since there are n pieces of each type available, there are $2(n)(n) = 2n^2$ distinguishable ways to do that (the factor of 2 is there because either part could go in either box). For the next two boxes, there are now $n-1$ pieces of each type available and so they can receive a long part and a short part in $2(n-1)^2$ distinguishable ways. Continuing in this manner, we see that

$$N_2 = \{2n^2\}\{2[(n-1)^2]\}\{2[n-2]^2\}\ldots\{2[3]^2\}\{2[2]^2\}\{2[1]^2\} = 2^n(n!)^2.$$

The answer to our problem is therefore $2^n(n!)^2/(2n)!$. There are 46 chromosomes in a human cell, so it is interesting to evaluate this expression for $n = 46$. You can see the denominator is 92!, which is a pretty big number (it overwhelms a small handheld calculator), and so using Stirling's asymptotic formula for $n!$ is helpful: $n! \sim \sqrt{2\pi n}\, n^n e^{-n}$. I'll let you confirm that our probability expression works out to be just about 1.7×10^{-13}, which is not very large.

The Twins

2.1 THE PROBLEM

In February 2008 I received a very interesting e-mail from Bruce C. Taylor, a professor of biomedical engineering at the University of Akron. Bruce had just been reading my book, *Duelling Idiots* (Princeton 2002), and that prompted him to write to me. Here's what Bruce wrote:

> I have an interesting probability problem that I have not been able to solve and I am just curious to see if you can come up with a solution. The problem came up when in one of our classes here I was assigning lab groups using a random number generator. As it turns out the class had 20 students, two of whom were related (twin sisters). Well, as luck would have it, the two sisters ended up in the same lab group of four. I had divided the class into five groups of four students. I, and a colleague, got to wondering what was the probability that the two sisters would end up in the same group. I originally thought that this would be a trivial problem but so far it has beaten me. I did write a MATLAB® program to solve the problem via a probabilistic model and I came up with a probability of 0.16 after 100,000 repetitions. I think that this is the correct answer but I can't, for the life of me, arrive anywhere near the same answer analytically. I thought maybe you'd like to take a crack at it.

Well, who could resist that?

After a bit of thought I did arrive at a theoretical result, a rational fraction approximately equal to 0.1579, and so I wrote back to Bruce to ask, "You said the [Monte Carlo] estimate was 0.16. Was it actually somewhat less?" Back came Bruce's response: "I ran the simulation three times at 100,000 reps. each and came up with the following: (1) 0.1591,

(2) 0.1570, (3) 0.1557." Not too bad an agreement with my fraction. I then wrote my own MATLAB® simulation code, ran it for ten million repetitions, and got an estimate of 0.1579092, an even better agreement with my theoretical fraction.

2.2   THEORETICAL ANALYSIS

To theoretically derive the answer to Bruce's question, here's what I sent him, where $\binom{x}{y}$ is, as in the first problem, the binomial coefficient $x!/(x-y)!y!$, with x and y both non-negative integers and $y \le x$.

First, to find the total number of ways (TNW) to randomly place 20 students into 5 groups of 4 each, imagine 5 bins. In the first bin we place 4 from 20, then 4 from the remaining 16 in the second bin, then 4 from the remaining 12 in the third bin, and so on. Thus, TNW $= \binom{20}{4}\binom{16}{4}\binom{12}{4}\binom{8}{4}\binom{4}{4}$.

Next, to find the total number of ways that the twins are together (TNWTT) in the same bin, we first imagine that the twins are glued together. When we select a twin, we automatically select the other one, too. There are 5 ways to place the glued twins into one of the bins, leaving 18 students. There are $\binom{18}{2}$ ways to select the 2 students who join the twins, leaving 16 students. We then finish the analysis as before, that is TNWTT $= 5\binom{18}{2}\binom{16}{4}\binom{12}{4}\binom{8}{4}\binom{4}{4}$. The probability we are after is

$$\frac{\text{TNWTT}}{\text{TNW}} = \frac{5\binom{18}{2}}{\binom{20}{4}} = \frac{5 \, 18!/16!2!}{20!/16!4!} = 5\frac{18!4!}{20!2!} = 5\frac{(4)(3)}{(20)(19)}$$

$$= \frac{3}{19} = 0.15789\ldots$$

Now, as easy as the above analysis may appear, an early reviewer of this book (Nick Hobson) pointed out to me that there is an even easier way to see the result in a flash. A total of 20 lab slots are to be filled, with 4 slots in each lab section. One of the twins, of course, has to be in *some* lab section, leaving 3 slots in *that* section still available out of the 19 total slots that are still available. So, the probability that our second twin gets one of those 3 slots (and so joins her sister) is 3/19. That's it!

2.3   COMPUTER SIMULATION

To write a Monte Carlo simulation, I found the following imagery helpful. (I wrote my simulation code before receiving Nick's clever observation, so perhaps there is a better way to simulate—I'll leave that for *you* to explore!) I started by visualizing the 20 students lined up in front of me in some (random) order, standing in a row, shoulder to shoulder. Each holds a slip of paper. These slips each have a single number on them; there's a 2 on each twin's slip, while all the other students have a 1 on their slips. Starting at the far left (student 1), the first four students are assigned to lab section 1, the next four students to lab section 2, and so on, with students 17 through 20 assigned to lab section 5. To simulate the placement of the twins into their lab sections, all we need do is randomly generate two different integers from 1 to 20, integers that determine the positions where the twins stand in the shoulder-to-shoulder row.

The simulation code can determine if the two twins have been assigned to the same lab section by simply adding up the numbers, in each lab section, on the paper slips held by the students in that section. If a lab section has neither twin, the group sum will be 4, while if a lab section has one twin, the group sum will be 5. A group sum of 6, however, means we have a lab section that contains both twins. This is the decision logic behind the simulation code **twins.m**. I make no claims that **twins.m** is a superoptimal (in some sense) code, just that it is easily understood and executes in a reasonably short time (ten million repetitions on my quite ordinary, bottom-of-the-line computer required less than 23 seconds to run). After the code listing, I'll give you a quick walkthrough of what each line is doing (the line numbers at the far left are not part of the code but are included simply as reference tags for the walkthrough).

---

twins.m

```
01   together=0;
02   for loop1=1:10000000
03      lab=ones(1,20);
```

(continued)

(continued)

```
04    twin1=floor(20*rand)+1;
05    twin2=twin1;
06    while twin1==twin2
07       twin2=floor(20*rand)+1;
08    end
09    lab(twin1)=2;
10    lab(twin2)=2;
11    groupsum=zeros(1,5);
12    for loop2=1:5
13       x=4*(loop2-1);
14       for loop3=1:4
15          groupsum(loop2)=groupsum(loop2)+lab(x+loop3);
16       end
17    end
18    for loop4=1:5
19       if groupsum(loop4)==6
20          together=together+1;
21       end
22    end
23  end
24  together/10000000
```

Line 01 initializes the variable *together* to zero; at the end of ten million simulations *together* will be the number of simulations in which the twins were assigned to the same lab section. Lines 02 and 23 define the outer *for/end* loop that cycles the code through the ten million simulations. Line 03 defines the row vector *lab*, with all of its 20 elements initially equal to 1. The value *lab(k)* is the number written on the slip of paper held by the student in the *k*th row position. Initially, then, all 20 students have a 1 on their individual slips of paper. Line 04 assigns *twin1* equal to an integer value selected at random from 1 to 20, and line 05 assigns the same integer to *twin2*. Since the two twins can't, of course, have the same position in *lab*, lines 06 through 08 then continually assign *twin2* a new random integer value until *twin1* and *twin2* have different integer values. Lines 09 and 10 write a 2 on the slip of paper each twin holds, leaving the other 18 students holding slips of paper each with a 1. Line 11 initializes all five elements of the row vector *groupsum* to zero. The two nested loops defined by lines 12 through 17 run through the 20 elements of *lab*, four at a time, from left to right, and generate the five element values of *groupsum*. Finally, the two nested loops defined by lines 18 through 22 check each element of *groupsum* and, if a value of 6 is detected (indicating both twins are in the same section), then *together* is incremented by one. Once the ten million simulations are finished, line 24 prints the code's estimate of the probability of the twins being in the same lab section (0.1579092), an estimate very close to the theoretical value.

Steve's Elevator Problem

3.1 THE PROBLEM

In my book *Digital Dice* (Princeton 2008), I included a puzzle question called "Steve's elevator problem." It was named for a reader in California, Steve Saiz, who had written to me about it after reading my book *Duelling Idiots* (Princeton 2002). As I wrote in *Digital Dice*, here's how Steve explained it to me in a March 2004 e-mail:

> Every day I ride up to the 15th floor in an elevator. This elevator only goes to floors G, 2 8, 9, 10, 11, 12, 13, 14, 15, 16, and 17. On average, I noticed that I usually make 2 or 3 stops when going from the ground floor, G, to 15, but it really depends on the number of riders in the elevator car. Is it possible to find the expected value for the number of stops the elevator makes during my ride up to the 15th floor, given the number of riders?

I was able, using combinatorial arguments, to derive expressions for the expected number of stops (in a building with n floors above floor G) if there are, initially on floor G, k riders in addition to Steve for the two cases of $k = 1$ and $k = 2$. Specifically, if $k = 1$, then the average number of stops is $2 - 1/n$, and if $k = 2$, then the average number of stops is $3 - 7/n + 3/n^2$. You can find these two derivations in *Digital Dice* on pp. 49–50 and pp. 124–125. In addition, another special case was solved, again using combinatorial arguments, by Michel Durinx at Leiden University in the Netherlands (for any value of k, assuming that $n = 11$), with the answer $9 - 8(10/11)^k$. (The amusing story of how Michel got into this discussion is told on pp. 127–128 of *Digital Dice*.)

The status of the problem them remained unchanged until I received, in October 2008, an e-mail from Shane G. Henderson, a professor in the School of Operations Research and Information Engineering at Cornell University. Shane wrote to tell me he had found a *complete, general, noncombinatorial* analytical solution to Steve's elevator problem!

3.2   THEORETICAL ANALYSIS

Shane's analysis is beautiful in its amazing simplicity. Define X as a random variable whose value equals the number of stops for Steve as his elevator starts (with k other people) its ride up from floor G toward floors $1, 2, \ldots, n-2$. In addition, let's also define the $n-3$ random variables I_i, $1 \leq i \leq n-3$, as follows:

$$I_i = \begin{array}{l} 1 \text{ if the evevator stops at floor } i, \\ 0 \text{ if the evevator does not stop at floor } i. \end{array}$$

(Mathematicians call random variables like this *indicator* functions, hence the I notation.) Then,

$$X = 1 + \sum_{i=1}^{n-3} I_i$$

where the 1 is the certain stop at floor $n-2$ (Steve's floor) and the summation runs only up to floor $n-3$ because Steve doesn't care what happens above floor $n-2$ (and the 1 has already accounted for the stop at floor $n-2$). Taking expectations,

$$E(X) = E\left[1 + \sum_{i=1}^{n-3} I_i\right] = 1 + \sum_{i=1}^{n-3} E(I_i).$$

We can calculate $E(I_i)$ as

$$E(I_i) = 0 \times \mathrm{Prob}\,(I_i = 0) + 1 \times \mathrm{Prob}\,(I_i = 1) = \mathrm{Prob}\,(I_i = 1),$$

and so

$$E(X) = 1 + \sum_{i=1}^{n-3} \mathrm{Prob}\,(I_i = 1) = 1 + \sum_{i=1}^{n-3} \mathrm{Prob}\,(\text{elevator stops at floor } i).$$

So far this is all pretty straightforward. Now for Shane's essential insight: the probability the elevator stops at any particular floor *is the same* as that for stopping at any other floor. If this was not so, then we'd be faced with the problem of explaining why not, that is, why would one floor (at the instant the elevator starts its run up the elevator shaft) be more or less likely to be a stop than some other floor? We have, in fact, no a priori reason to believe that is the case.

As the elevator approaches floor i, $1 \le i \le n-3$, there will be on-board a number of riders each of whom might get off on that floor (but not Steve, who *will* get off *only* at floor $n-2$), a number that can range from zero (because all of Steve's companions have already got off on lower floors) to k (because nobody has yet got off). So, in general, the calculation of the probability that the elevator stops at floor i involves a conditional probability analysis, that is,

$$\begin{aligned} &\mathrm{Prob}(\text{elevator stops at floor } i), \quad 1 \le i \le n-3, = \\ &\sum_{l=0}^{k} \mathrm{Prob}(\text{elevator stops at floor } i \mid l \text{ riders who might get off}) \\ &\qquad \times \mathrm{Prob}(l \text{ riders who might get off}). \end{aligned}$$

This calculation is not horrible to do, but we can avoid it completely by simply noticing that, for the particular case of $i = 1$, we are guaranteed that all $k+1$ original riders (Steve, plus his k initial companions) are still in the elevator (and so up to k people could get off) because nobody has yet had a chance to get off. This special case calculation is easy to do:

$$\begin{aligned} \mathrm{Prob}(\text{elevator stops at floor } 1) &= 1 - \mathrm{Prob}(\text{elevator does not stop} \\ \text{at floor } 1) &= 1 - \mathrm{Prob}(\text{nobody gets off at floor } 1). \end{aligned}$$

We, of course, know Steve is not going to get off at floor 1. That much is certain, and so Prob(nobody gets off at floor 1) = Prob(none of the k additional riders gets off at floor 1), a probability that we can write by inspection simply by noticing that any particular one of these k riders gets off on floor 1 with probability $1/n$ and so does *not* get off on floor 1 with probability $1 - 1/n$. Thus,

Prob(nobody gets off at floor 1) $= (1 - 1/n)^k$, and therefore
Prob(elevator stops at floor 1) $= 1 - (1 - 1/n)^k$.

This immediately tells us that

$$E(X) = 1 + \sum_{i=1}^{n-3}\left\{1 - \left(1 - \frac{1}{n}\right)^k\right\}$$

and so, at last, we have Shane's elegant result:

$$E(X) = 1 + (n-3)\left\{1 - \left(1 - \frac{1}{n}\right)^k\right\}.$$

We can partially check this general formula against the three special cases I mentioned earlier.

Case 1: For $n = 11$ and all possible k (the original "Steve's elevator problem"),

$$E(X) = 1 + 8\left[1 - \left(1 - \frac{1}{11}\right)^k\right] = 1 + 8 - 8\left(\frac{10}{11}\right)^k$$
$$= 9 - 8\left(\frac{10}{11}\right)^k,$$

the result found, using fairly complicated combinatorial arguments, by Michel Durinx (see p. 127 in *Digital Dice*).

Case 2: For $k = 1$ and all possible n,

$$E(X) = 1 + (n-3)\left[1 - \left(1 - \frac{1}{n}\right)\right] = 1 + \frac{n-3}{n}$$
$$= 2 - \frac{3}{n}$$

the result found, using simple combinatorial arguments, by me (see pp. 49–50 in *Digital Dice*).

Case 3: For $k = 2$ and all possible n,

$$E(X) = 1 + (n-3)\left[1 - \left(1 - \frac{1}{n}\right)^2\right]$$

$$= 1 + (n-3)\left[1 - 1 + \frac{2}{n} - \frac{1}{n^2}\right]$$

$$= 1 + (n-3)\left[\frac{2}{n} - \frac{1}{n^2}\right]$$

$$= 1 + \frac{2(n-3)}{n} - \frac{n-3}{n^2} = 1 + 2 - \frac{6}{n} - \frac{1}{n} + \frac{3}{n^2}$$

$$= 3 - \frac{7}{n} + \frac{3}{n^2},$$

the result found, using slightly more complicated combinatorial arguments, by me (see pp. 124–125 in *Digital Dice*).

Notice that Shane's analysis is easily modified to handle the possibility that Steve's office is moved to some other floor. A brute force combinatorial approach, on the other hand, would be far more disrupted by such a change.

3.3   COMPUTER SIMULATION

In *Digital Dice* I included a simulation code called **steve.m** on p. 126, and I'll not repeat it here. (You can find it on Princeton University Press's math website entry for the book.) That code was for the specific case of $n = 11$ (a value easily changed) and for any value of k.

Three Gambling Problems Newton Would "Probably" Have Liked

4.1 THE PROBLEMS

To lay the groundwork for the first problem, suppose you give a fair die to each of a large number of people. Each person tosses his or her die, over and over, until a 6 appears. Once this has occurred for every person, each then reports on how many tosses he or she made. What do you think is the average number of tosses of a fair die to get the first 6? Since the probability of a 6 on a fair die is 1/6, most people immediately reply "six tosses," whether or not they really understand why. In fact, the answer is easy to calculate, and six is indeed correct. Here's how to do it (but note, this is not one of the three "Newton" problems we are going to solve in this puzzler, merely a warm-up for the ones we'll really be interested in, in just a few moments).

When a die is tossed and a 6 appears, let's call that a *success* and write it as S. If anything else appears we'll call that a *failure* and write it as F. Mathematicians call sequences representing repeated trials with just two possible, independent outcomes on each trial—our S's and F's—*Bernoulli trials*, in honor of the Swiss mathematician James Bernoulli, mentioned in note 9 of the introduction. Now, let's write $\text{Prob}(S) = p$ and $\text{Prob}(F) = 1 - p$. We can write down all the possible sequences of S's and F's that represent the appearance of the first S as follows: S, FS, FFS, FFFS, and so on. That is, if n is the number of tosses until the first 6 appears (the first S), then we have $\text{Prob}(n = 1) = p$, $\text{Prob}(n = 2) = (1 - p) p$, $\text{Prob}(n = 3) = (1 - p)^2 p$, $\text{Prob}(n = 4) = (1 - p)^3 p$ and, in general, $\text{Prob}(n = k) = (1 - p)^{k-1} p$.

Now, what we want to calculate is the average value of n, formally written as

$$\mu = \sum_{n=1}^{\infty} nP(n).$$

We could directly evaluate this sum, but that's not the approach I'll use. Instead, I'm going to show you what might at first seem to be just a trick. But, as you'll see when we get to the first "Newton" problem, the same trick will work on it, too, and as is well known in mathematics, any trick that you can use more than once is actually a method!

We start by observing the obvious: any sequence of S's and F's starts either with an S or with an F. If it is with an S (with probability p), then we have the average number of tosses until the first S as 1. If it is with an F (with probability $1 - p$), then we have the average number of tosses until the first S as $1 + \mu$. So, weighting these two particular results for the average number of tosses until the first S with their probabilities, we have

$$\mu = 1(p) + (1 + \mu)(1 - p) = p + 1 + \mu - p - \mu p = 1 + \mu - \mu p,$$

or

$$0 = 1 - \mu p,$$

and so, at last (and with surprising suddenness!),

$$\mu = \frac{1}{p},$$

which is 6 for a fair die, just as most people intuitively guess.

Now you are ready for the first Newton puzzle in this problem. What is the average number of tosses of a fair die until the first time two 6s consecutively appear? This is a more difficult problem than the warm-up one because there now appears to be no obvious pattern to the sequences of S's and F's that end for the first time in SS. (Try it and see!) If you think intuition might still work here and say 12 tosses (perhaps because $6 + 6 = 12$) or 36 tosses (perhaps because $6^2 = 36$)—well, neither guess is correct. The correct answer is larger than even the 36. What we need to do is some more analysis rather than mere guessing.

The title of this chapter says there are *three* gambling problems Newton would probably have liked. What are the other two? The second one is very easy to state, but maybe not so easy to get your head around. It would, I think, have given even the great Newton pause for thought. Given a coin with probability p for heads on a flip and probability $1 - p$ for tails, what is the probability that, if we flip the coin over and over, we'll get the fourth head before the seventh tail? What is the numerical value of this probability if the coin is fair ($p = 1/2$)? What is the probability if the coin is "almost" fair ($p = 0.45$)?

Finally, the third Newton gambling problem is just as easy to state and at least as tricky to get a handle on. Three people—call them A, B, and C—take turns tossing a single fair die. A starts, followed by B, then C, then back to A, and so on. This continues until one of them tosses a 6. That player then drops out of the game, and the remaining two continue. Eventually one of them finally tosses a 6, and the game ends. What is the probability it is first A, then B, that first toss 6s? What is the probability it is first B, then A, that first toss 6s?

4.2   THEORETICAL ANALYSIS 1

For the first Newton problem, let's write $P(n)$ for the probability our die first shows two 6s in a row at the completion of n tosses. Obviously, $P(1) = 0$ and $P(2) = p^2$. We can even do a couple more cases by hand simply by writing out all possible sequences of S and F of length n and observing which ones end with SS and contain no earlier consecutive S's. For example, $P(3) = (1 - p) p^2$ for FSS and $P(4) = (1 - p)^2 p^2 + (1 - p) p^3$ for FFSS plus SFSS. For longer sequences, however, this enumeration approach becomes increasingly difficult to do. We need a different approach. Let's try our trick again!

Now we begin by observing that any sequence of S's and F's starts with one of four possibilities: SS, FF, SF, or FS. If it starts with SS (with probability p^2), then we have the average number of tosses until the first double S as 2. If it starts with either FF or SF (with probability $(1 - p)^2 + p(1 - p)$), then we have the average number of tosses until the first double S as $2 + \mu$. And finally, if it starts with FS (with probability $(1 - p)p$), then with a following S we have the average number of tosses until the first double S as 3 (that is, with probability $(1 - p)p^2$), or with a following F we have the average number of tosses until the first double S as $3 + \mu$ (that is, with probability $(1 - p)^2 p$). And so, weighting the individual averages with their probabilities,

$$\mu = (2)\, p^2 + (2 + \mu)\, [(1 - p)^2 + p(1 - p)] + (3)\, (1 - p)\, p^2 + (3 + \mu)\, [(1 - p)^2 p].$$

If you go through the algebra you should arrive at

$$\mu = \frac{2 + p - p^2}{p^2(2 - p)} = \frac{2 - p + 2p - p^2}{p^2(2 - p)},$$

or, at last,

$$\mu = \frac{1}{p^2} + \frac{1}{p}.$$

That is, for a fair die, $\mu = 42$ tosses, a number I'm pretty sure nobody (or, at least, hardly anybody) would have guessed!

4.3   COMPUTER SIMULATION 1

The code **ds.m** (for double S) simulates the repeated tossing of a fair die until the first double 6 appears. Indeed, if we call a sequence of such tosses a game, then **ds.m** simulates 100,000 games, all the while keeping track of how many tosses occur in each game until the first double 6 appears. Finally, it averages those 100,000 numbers to estimate the answer to our question. When **ds.m** was run, it produced an estimate of 41.8865 tosses, which is pretty close to our theoretical result. Here's how the code works.

The first command calculates the probability of tossing a success (a 6) with a fair die and calls it *check*. Then, to simulate a game (the current game number is the value of the main loop variable *loop*, which runs from 1 to 100,000), the variables *toss* and *flip2* are first initialized. The variable *toss* is the number of tosses that the toss about to be performed will be in the current game, and so *toss* is initially set equal to 1. A toss is performed by setting *result* equal to a random value between 0 and 1 and, if *result* is less than *check*, we have an S (and *flip1* is set equal to 1)— otherwise we have an F and *flip1* is set equal to 0. Since the code only has to remember the outcomes of the two most recent tosses, *flip2*, which represents the outcome of the previous toss, is initially set equal to 0.

To simulate a game, a *while* loop operates as long as *keeptossing* is equal to 1 (and so *keeptossing* is initially set equal to 1). Every time a new toss is about to be made, the old value of *flip1* is given to *flip2*, and then *flip1* gets the new outcome. To see if a toss has been the second S in a row, the code asks if *flip1* + *flip2* = 2; if so, *keeptossing* is set to zero, which then causes the *while* loop to terminate. The vector *person* stores the present value of *toss*: *person(j)* is the number of tosses in the *j*th game until the first double S. If *flip1* + *flip2* ≠ 2, then the code tosses again: *toss* is incremented by 1 and *flip2* "remembers" the value of *flip1*. Then, the *while* loop operates again. When all 100,000 games are finally completed, the last command of **ds.m** sums the 100,000 values in the *person* vector and prints their average value.

---

ds.m
```
check=1/6;
for loop=1:100000
   toss=1;flip2=0;
   keeptossing=1;
   while keeptossing==1
      result=rand;
      if result<check
         flip1=1;
      else
         flip1=0;
      end
      if flip2+flip1==2
         keeptossing=0;
         person(loop)=toss;
      else
```

(continued)

(continued)

```
        toss=toss+1;
        flip2=flip1;
    end
  end
end
sum(person)/100000
```

4.4 THEORETICAL ANALYSIS 2

What makes the second of our two problems hard is that it is not at all obvious (at first, anyway) where to start. If we make a couple of simple observations, however, and put them together, the solution is surprisingly easy. First, notice that the event of interest *will* occur by the end of the tenth flip. It is certain to happen either when we get the fourth head on that flip after six earlier flips gave tails or we get the seventh tail. The event could, of course, occur in fewer flips, perhaps as soon as on the fourth flip (with four straight heads). This gives us a clue as to where to start our analysis.

What we'll do is calculate the probabilities that we get the fourth H on the tenth flip, that we get the fourth H on the ninth flip, that we get the fourth H on the eighth flip, and so on, down to getting the fourth H on the fourth flip. Our final answer will be the sum of all these probabilities. To do this it will be very helpful to have a systematic way of writing down these individual probabilities. Here's how to do that. The probability of getting *exactly* four H's in exactly n flips means that we get the fourth H on the nth flip *and that there were three earlier H's somewhere in the previous $n - 1$ flips*. That can happen in $\binom{n-1}{3}$ ways, with each way having probability $p^3(1-p)^{(n-1)-3} = p^3(1-p)^{n-4}$, because if there were three H's in $n-1$ flips, then there must have been $(n-1) - 3 = n - 4$ T's in those $n-1$ flips. The final, nth flip gives us the fourth H (with probability p), and so the probability of exactly four H's in exactly n flips is $\binom{n-1}{3} p^4 (1-p)^{n-4}$. The answer to our question is then

$$\sum_{n=4}^{10}\binom{n-1}{3}p^4(1-p)^{n-4} = p^4\sum_{n=4}^{10}\binom{n-1}{3}(1-p)^{n-4}$$

$$= p^4\left[\begin{array}{c}1 + \binom{4}{3}(1-p) + \binom{5}{3}(1-p)^2 + \binom{6}{3}(1-p)^3 + \binom{7}{3}(1-p)^4 \\ + \binom{8}{3}(1-p)^5 + \binom{9}{3}(1-p)^6\end{array}\right]$$

$$= p^4\left[\begin{array}{c}1 + 4(1-p) + 10(1-p)^2 + 20(1-p)^3 + 35(1-p)^4 \\ + 56(1-p)^5 + 84(1-p)^6\end{array}\right].$$

If $p = 1/2$, then this all reduces to $848/1{,}024 = 0.828125$. If $p = 0.45$, the result is that the probability of getting four heads before the seventh tail is 0.733962.

4.5   COMPUTER SIMULATION 2

The code **before.m** simulates, ten million times, our coin-flipping problem. Here's how it works. After setting the value of p, the variable *total* is initialized to zero. At the end of ten million simulations of our random process, *total* will be the number of those simulations in which four heads occurs before seven tails do. Each simulation begins with H (the number of heads, so far) and T (the number of tails, so far) set to zero. Then, a *while* loop executes until either H reaches 4 or T reaches 7, with either condition terminating the loop. If it was the H = 4 condition, then *total* is incremented by one. Once all the simulations are completed, the last line of the code produces an estimate for our probability.

---

before.m
```
p=0.45;total=0;
for loop=1:10000000
    H=0;T=0;
    while H<4&&T<7
        if rand<p
```

(continued)

(continued)
```
        H=H+1;
        else
            T=T+1;
        end
    end
    if H==4
        total=total+1;
    end
end
total/10000000
```

When run, **before.m** produced estimates of 0.827948 for $p = 1/2$ and 0.734308 for $p = 0.45$; both estimates are close to the theoretical values we calculated earlier.

4.6   THEORETICAL ANALYSIS 3

Let P_A be the probability that A is the first to toss a 6 as A, B, and C take turns tossing a fair die. Then either A tosses his 6 on his first try (with probability $1/6$) or he and the other two players do *not* toss a 6 (with probability $(5/6)^3$, and we are back to our original starting condition. So,

$$P_A = \frac{1}{6} + \left(\frac{5}{6}\right)^3 P_A,$$

which is easily solved to give

$$P_A = \frac{36}{91}.$$

With this probability A leaves the game, and B and C continue. With exactly the same sort of argument, if P_B is the probability B is the first between B and C to toss a 6, we then have

$$P_B = \frac{1}{6} + \left(\frac{5}{6}\right)^2 P_B,$$

which is easily solved to give

$$P_B = \frac{6}{11}.$$

Thus, the probability it is A, and then B, to first toss 6s is

$$\left(\frac{36}{91}\right)\left(\frac{6}{11}\right) = \frac{216}{1,001} = 0.21578\ldots$$

For the reverse of the above, that is, for the case it is B, then A, that are the first to toss 6s, let's take a different approach. Notice that the sequence of play is described by

$$ABCABCABCABCA\ldots,$$

and so we see that

$$P_B = \left(\frac{5}{6}\right)\left(\frac{1}{6}\right) + \left(\frac{5}{6}\right)^4\left(\frac{1}{6}\right) + \left(\frac{5}{6}\right)^7\left(\frac{1}{6}\right) + \cdots$$
$$= \left(\frac{5}{6}\right)\left(\frac{1}{6}\right)\left[1 + \left(\frac{5}{6}\right)^3 + \left(\frac{5}{6}\right)^6 + \cdots\right].$$

The geometric series in the brackets on the right has the sum

$$\frac{1}{1 - \left(\frac{5}{6}\right)^3},$$

and so

$$P_B = \frac{\left(\frac{5}{6}\right)\left(\frac{1}{6}\right)}{1 - \left(\frac{5}{6}\right)^3} = \frac{(5)(6)}{6^3 - 5^3} = \frac{30}{91}.$$

With this probability, B leaves the game, and C and A continue, with C having the next toss. The alternating sequence of play is now described by

$$CACACAC\ldots,$$

and so we see that

$$P_A = \frac{\left(\frac{5}{6}\right)\left(\frac{1}{6}\right)}{1-\left(\frac{5}{6}\right)^2} = \frac{5}{6^2-5^2} = \frac{5}{11}.$$

Thus, the probability that B, and then A, first toss 6s is

$$\left(\frac{30}{91}\right)\left(\frac{5}{11}\right) = \frac{150}{1,001} = 0.14985\ldots.$$

Big Quotients—Part 1

5.1 THE PROBLEM

This is the first part of a two-part problem and so, of course, this will be the easy part. But don't skip it! Understanding what we do here will be a big help in understanding the second part. Here's the question: pick a number at random (uniformly), from 0 to 1. Pick another number, also at random (uniformly), from 0 to 1. What is the probability that if you divide the larger number by the smaller one the answer will be greater than 2? Greater than 3? More generally, greater than k, where $k \geq 1$?

Before I show you how to answer these initial questions, let me tantalize you with a generalization of them that we'll tackle in "Big Quotients—Part 2" later in the book (Problem 16). Pick N numbers, each independently and randomly (uniformly), from 0 to 1. What is the probability that if you divide the largest of these N numbers by the smallest, the answer will be greater than k, where $k \geq 1$? The original question is obviously the $N = 2$ special case of this generalized version; if we can get the answer to the first question, then we can use it as a check on the answer to the generalized question.

5.2 THEORETICAL ANALYSIS

Let's call our two numbers, each picked at random from 0 to 1, X_1 and X_2. We don't know which is the larger, so if we agree to always calculate X_2/X_1, then we are interested in both the probability $X_2/X_1 > k$ (which accounts for the case $X_2 > X_1$) and the probability $X_2/X_1 < 1/k$ (which

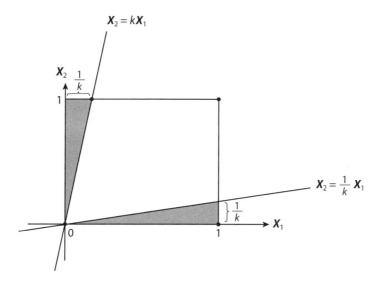

Figure 5.2.1. The *unshaded* area is where $1/k \le X_2/X_1 \le k$.

accounts for the case $X_2 < X_1$). A more elegant way to express this is with a double inequality, as $1 - \text{Prob}(1/k \le X_2/X_1 \le k)$, $k \ge 1$.

In Figure 5.2.1 I've drawn a two-axis coordinate system, with X_2 on the vertical axis and X_1 on the horizontal axis. Each point in the unit square in the first quadrant represents a way that X_1 and X_2 can take on their values, and the unit area inside that square represents the totality of all possible values for X_1 and X_2. That area of the square is the *sample space* for our problem. The entire unit area, therefore, has probability 1 (some point in that area, with coordinates X_2, X_1 is certain to occur). The central, unshaded area in the square is the area associated with the double inequality $1/k \le X_2/X_1 \le k$, because that is the area containing all the points such that $X_2 \le kX_1$ *and* $X_2 \ge (1/k) X_1$ (that is, all the points both below the line $X_2 = kX_1$ and above the line $X_2 = (1/k) X_1$ are in the unshaded area).

The probability associated with the unshaded area is just that area itself, because X_1 and X_2 are *uniformly distributed* (this is the fundamental assumption behind what is called *geometric probability*). The probability we are after is the probability that we are not in the unshaded area but rather in the shaded area, and that probability is the total area of the shaded regions. By elementary geometry, that area is $2(1/2)(1)$ $(1/k) = 1/k$. Okay, that's it, we are done! Our answer is

$$\text{Prob}(\text{"larger number divided by smaller number} > k\text{"}) = \frac{1}{k}.$$

5.3   COMPUTER SIMULATION

The code **ratio1.m** simulates ten million selections of X_1 and X_2, and keeps track (in total) of how many times the larger divided by the smaller gives a result (r, for ratio) greater than 2. If we simply change the test on r to different values of k, we can easily check our theoretical result, as shown in Table 5.3.1, with pretty good agreement.

```
ratio1.m
total=0;
for loop=1:10000000
  x1=rand;x2=rand;
  k=2;
  r=x2/x1;
  if r>k|r<1/k
     total=total+1;
  end
end
total/10000000
```

Table 5.3.1. Theory versus 'experimentation' (N = 2)

k	theory	simulation
2	0.5000000	0.4997159
3	0.3333300	0.3333913
4	0.2500000	0.2500693
5.5	0.1818181	0.1817171

An easy modification to **ratio1.m** lets us simulate the second question, too, so let's do that before we eventually do its theoretical analysis in Problem 16. The new code, **ratio2.m**, is for the next step up in complexity situation, when we pick $N=3$ numbers independently at random (uniformly) from 0 to 1, and divide the largest by the smallest.

```
ratio2.m
total=0;
for loop=1:10000000
    for j=1:3
        x(j)=rand;
    end
    k=2;
    r=max(x)/min(x);
    if r>k|r<1/k
        total=total+1;
    end
end
total/10000000
```

The results for the same values of k that I used in Table 5.3.1 are shown for this situation in Table 5.3.2. You can see there is a significant difference in the $N=3$ case compared to the $N=2$ case. To see if these results agree with theory, we need to do some more theory! We'll come back to this in Problem 16.

Table 5.3.2. Just 'experimentation'
at this point in the book (N = 3)

k	theory	simulation
2	?	0.7498262
3	?	0.5553961
4	?	0.4372493
5.5	?	0.3306806

Two Ways to Proofread

6.1 THE PROBLEM

As anyone who has ever written a book can tell you, one of the most important tasks—but a tedious one—is proofreading the initial type-set pages. There are *always* misprints! The only way to catch and correct them is through careful reading of the entire book, a job that is all too easy to nod off on while doing. There are at least two ways to do it, where I'll make the following two assumptions: (1) there is an unknown total of m misprints in the book, and (2) the probability a reader spots a misprint, given that he is looking at a misprint, is a constant (ideally, we'd hope that probability is 1, but that's probably not very realistic).

Method 1: Two readers independently examine the page proofs. Reader 1 flags a misprints and Reader 2 flags b misprints. There are c flagged misprints in common. Assume Reader 1 has probability p of spotting a misprint, and Reader 2 has probability q. We know p and q from the past performance of each reader.

Method 2: One reader reads the proofs at two different times, times sufficiently separated that we can assume the two readings are independent. On the first reading he flags a misprints and circles them in red so he'll know he's already found them when he does the second reading. On the second reading he spots b additional misprints.

For each method, find an expression for U, the number of remaining, undetected misprints. Evaluate your two expressions for these values: for Method 1, $a = 30$, $b = 25$, and $c = 5$, and for Method 2, $a = 30$ and $b = 20$.

6.2   THEORETICAL ANALYSIS

For Method 1, we can write

$$mp = a, \ mq = b, \ mpq = c.$$

The third equation follows by reasoning that mp is the number of misprints found by Reader 1, and it is from that subset of misprints that Reader 2 must find the misprints that both readers find in common. So,

$$p = \frac{a}{m}$$

and

$$q = \frac{b}{m}.$$

Thus,

$$m\left(\frac{a}{m}\right)\left(\frac{b}{m}\right) = c = \frac{ab}{m},$$

and therefore

$$m = \frac{ab}{c}.$$

Since the total number of different misprints spotted is $a + b - c$, the total number of undetected misprints is given by

$$U_1 = m - (a + b + c) = \frac{ab}{c} - a - b + c = \frac{ab - ac - bc + c^2}{c},$$

or, for Method 1,

$$U_1 = \frac{(a-c)(b-c)}{c}.$$

Notice carefully that there is no explicit dependence in U_1 on either p or q, but rather the influence of these two probabilities is in the particular values of a, b, and c that occur. For the values given,

$$U_1 = \frac{(30-5)(25-5)}{5} = 100.$$

For Method 2, we can write

$$mp = a$$

and

$$(m - a)p = b.$$

So,

$$p = \frac{a}{m},$$

and thus

$$(m - a)\frac{a}{m} = b,$$

or

$$ma - a^2 = mb.$$

This quickly gives us

$$ma - mb = a^2 = m(a - b)$$

and therefore

$$m = \frac{a^2}{a-b}.$$

Since the total number of misprints spotted is $a + b$, the number of undetected misprints is given by

$$U_2 = m - (a+b) = \frac{a^2}{a-b} - (a+b) = \frac{a^2 - (a+b)(a-b)}{a-b} = \frac{a^2 - (a^2 - b^2)}{a-b},$$

or

$$U_2 = \frac{b^2}{a-b}.$$

For the values given,

$$U_2 = \frac{400}{30-20} = 40.$$

The numerical results for U_1 and U_1 are pretty discouraging! Writing a zero-error book appears to be a very difficult task. I suspect, in fact, that it's pretty much a practical impossibility.

Chain Letters That Never End

7.1 THE PROBLEM

Just to be sure you don't think I am claiming computer simulation is a complete substitution for theory, here's an example of a random process that would be nontrivial to simulate, and we'll count ourselves lucky to have a theoretical solution. Suppose somebody decides to create a chain letter, and starts things off by sending a copy of it to C people. Each of those C people is asked to then send off C copies of their own, and so on. With probability p, any person who receives such a letter decides to ignore it. What is the probability that this chain letter eventually disappears? There is no guarantee that it will, after all, and simulating a process that potentially has no termination point presents some practical difficulties! Fortunately, there is a very nice way to attack this problem analytically.

7.2 THEORETICAL ANALYSIS

How does our chain letter die? For that to happen, each of the original C recipients can be thought of as starting a sub-chain letter sequence that eventually terminates with probability E (for *extinction*). There are just two ways such a sub-chain letter sequence can terminate: (1) with probability p the original recipient simply doesn't send his C copies, and (2) with probability $1 - p$ he *does* send his C copies, and then later all the resulting C sub-sub-chain letter sequences initiated by those copies eventually terminate (which happens with probability E^{C}). Thus, we arrive at the following C-order algebraic equation for E:

$$E = p + (1 - p) E^C.$$

Once we select a value for C, we can solve this equation for the extinction probability for a sub-chain, E, over and over, as we let p vary from 0 to 1. The value of E^C as p varies gives us the probability we are after, the probability that *all* of the sub-chains started by the original recipients terminate. The complementary probability, $1 - E^C$, is the probability that the chain letter goes on forever.

The equation has the obvious solution $E = 1$ for any value of C (and all p), but for interesting values of C (say, 2 to 5), there are other solutions as well (they include those that are negative, greater than 1, or complex, and so none can be a valid probability), and so we are faced with a fairly substantial number-crunching task. Fortunately, modern computer software is up to the job (I used MATLAB®'s powerful command *solve*), and Figures 7.2.1 through 7.2.4 show the probability E^C versus p for $C = 2$, 3, 4, and 5, respectively. For the case of $p = 1/2$ (each recipient flips a fair coin to decide whether to send his C copies or to forget it), we have, for example, $E^3 = 0.2361$ and $E^4 = 0.0874$. Adding just one additional copy (from $C = 3$ to $C = 4$) greatly reduces (by nearly a factor of 3) the probability the chain letter will eventually terminate.

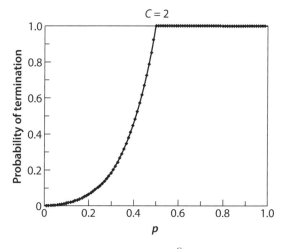

Figure 7.2.1. The probability E^C versus p for $C = 2$.

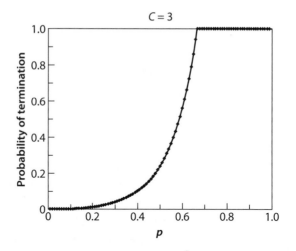

Figures 7.2.2. The probability E^C versus p for C = 3.

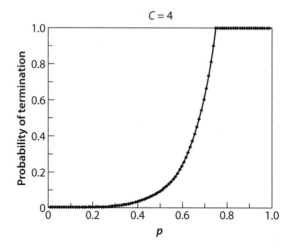

Figures 7.2.3. The probability E^C versus p for C = 4.

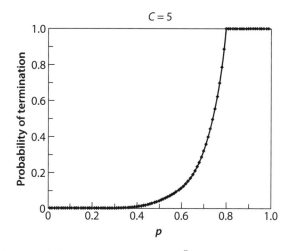

Figures 7.2.4. The probability E^C versus p for C = 5.

Bingo Befuddlement

8.1 THE PROBLEM

In the introduction (I.7), I discussed nontransitive dice. Nontransitivity can also occur in many other situations, far removed from dice. A game that kids love to play (some adults, too!), rock, paper, scissors, is nontransitive because while the rock breaks the scissors and the scissors cuts the paper, the paper covers the rock. For a more analytical example, imagine that we have nine tennis players ranked 1, 2, 3, 4, 5, 6, 7, 8, and 9 in increasing order of skill (8 *always* beats players 1 through 7, and always loses to player 9). Imagine further that these nine players form three teams A, B, and C, each with three players, as shown below:

Team A:	8	1	6
Team B:	3	5	7
Team C:	4	9	2

(It is interesting to note in passing that this particular structure is the famous Lo Shu, the unique three-by-three normal magic square that was known to the Chinese thousands of years before Christ [*normal* means the square is formed by using consecutive integers, starting with 1]. As it happens, this has absolutely nothing to do with what I am about to show you; it's simply a historical aside that I couldn't resist mentioning.)

These three teams, A, B, and C, engage in a series of matches, where a *match* means each player on a team plays one game with all the players on the opposing team. That is, a match consists of nine games. Now, notice that A wins a match with B by a score of 5 to 4 (8 wins all three of his games, 1 never wins, and 6 wins two games). Also, B wins a match with C, also by a score of 5 to 4 (3 wins one game, 5 wins two games, and 7 wins two games). So, A beats B and B beats C. Does that mean that A beats C? No, because C beats A by a score of 5 to 4 (4 wins one game, 9 wins three games, and 2 wins one game). The match results are not transitive. Very surprising, I think, but the numbers are beyond reproach.

In this new puzzle you'll see yet another case of nontransitivity that will be much more challenging to solve. In the cases of the dice and the tennis team problems we could study the interactions involved literally by counting. In this new problem there are so many possibilities that, unless you are a very patient person, you'd go quietly insane trying to do the arithmetic by hand. (Try it and see!) Here we'll find a computer to be our salvation. Another interesting twist to what is a probabilistic problem is that the computer analysis will not produce probabilistic estimates but rather will give us exact results! Here's the problem.

We start with four two-by-two bingo cards, A, B, C, and D. Two players each select adjacent cards, and then the numbers 1 through 6 are called in random order, one after the other, without repetition. That is, after six calls the numbers are exhausted. The first player to complete a horizontal row on his card wins. The claim here is that, on average, A beats B, B beats C, C beats D, and—here's the big claim—if we complete the loop, then D beats A. The problem is to confirm this claim, and to find the probabilities by which the winning cards beat the losing cards. It's important to realize that ties are possible (both players yell "Bingo!" simultaneously). For example, suppose cards A and B are being played. Then all sequences of called numbers that start with 1, 4, 2 result in a tie. Or if cards B and C are being played, all sequences of called numbers that start with 2, 5, 4 result in a tie. So don't forget about the possibility of ties.

$$A = \begin{bmatrix} 1 & 2 \\ 3 & 4 \end{bmatrix} \quad B = \begin{bmatrix} 2 & 4 \\ 5 & 6 \end{bmatrix} \quad C = \begin{bmatrix} 1 & 3 \\ 4 & 5 \end{bmatrix} \quad D = \begin{bmatrix} 1 & 5 \\ 2 & 6 \end{bmatrix}$$

8.2 COMPUTER SIMULATION

Since the six numbers 1 through 6 are called without repetition, there is only a finite number of possible sequences; specifically, there are $6! = 720$ sequences. Although finite, this is far too many to allow a reasonable hand calculation of how each pair of cards plays for each possible sequence. but it's a simple task for a computer. The MATLAB® code **bingo.m** does just that, playing each and all of the 720 possible sequences for any two given cards, all the while keeping track of which card wins on each sequence (or if a tie occurs).

When run for the given pairs, the code determined the probabilities we are after:

$$P(\text{A beats B}) = P(\text{B beats C}) = P(\text{C beats D}) = P(\text{D beats A}) = \frac{336}{720} = \frac{14}{30},$$

while the reverse probabilities are

$$P(\text{B beats A}) = P(\text{C beats B}) = P(\text{D beats C}) = P(\text{A beats D}) = \frac{312}{720} = \frac{13}{30},$$

and that for each of the card pairings,

$$P(\text{tie}) = \frac{72}{720} = \frac{3}{30}.$$

Here's a walkthrough on how **bingo.m** works. First I'll give a brief description of the general approach. The code plays cards X and Y against each other, where X and Y are each set equal to one of the cards A, B, C, or D (the particular choice is yours). As the numbers of a sequence are called, each sub-square in the X and Y cards is examined to see if it matches the current called number, and if it does, the sub-square entry is set equal to zero. The code "knows" when a card achieves bingo when one of the sums of the card's rows equals zero (this approach uses the advanced mathematical fact that $0 + 0 = 0$). If that happens for just *one* of X and Y, then that card wins. If *both* X and Y achieve a zero-sum row at the same time, we have a tie. A win or a tie is certain to occur by the end of any sequence. The code then restores X and Y to their initial card values (which were destroyed as sub-square entries were set to zero), and a new sequence is played.

Now, here's a bit more detail. The first two lines of the code define A, B, C, and D. The next line uses MATLAB®'s wonderful command *perms* to generate a matrix of all the permutations of the components of the vector that is the argument of the command. That is, *P* is a 720-by-6 matrix whose 720 rows are the permutations of the integers 1 through 6.

The code then sets *x* (the number of times card X wins) to zero, *y* (the number of times card Y wins) to zero, and ties (the number of times X and Y tie) to zero. The code then begins to loop through each of the rows of *P*, using the entries of a row as the current calling sequence, as follows.

After setting X = D (or X to any of the other possible cards) and Y = A (or Y to any of the other possible cards), *row* is set equal to the next available row of *P*, *xbingo* and *ybingo* are both set equal to zero (*xbingo* = 1, for example, means X has achieved bingo), and *keepplaying* is set equal to 1. The variable *i* is set equal to 1, which will step the code through the components of *row* to generate the calling sequence.

A *while* loop is then entered and, as long as *keepplaying* remains equal to 1, the components of *row* are called, one after the other; each time the appropriate sub-square of X and/or Y is set to 0. Then the row-sums of X and Y are calculated and, if a zero sum is found, either (or both) of *xbingo* and *ybingo* is (are) set equal to 1. (If neither is set equal to 1, then *i* is increased by one and the next component in *row* is called.) If both are equal to 1, then we have a tie (and *tie* is increased by one). If only one is equal to 1, then we have a win (and either *x* or *y* is increased by one). In any case, if bingo has occurred, then *keepplaying* is set equal to 0, which then causes the code to exit the *while* loop. This reinitializes the code variables (in particular, X and Y), and the next row of *P* is played.

---

bingo.m
```
A=[1,2;3,4];B=[2,4;5,6];
C=[1,3;4,5];D=[1,5;2,6];
P=perms([1,2,3,4,5,6]);
x=0;y=0;ties=0;
for loop=1:720
    X=D;Y=A;
    row=P(loop,:);
    keepplaying=1;i=1;xbingo=0;ybingo=0;
    while keepplaying==1
        n=row(i);
        for j=1:2
            for k=1:2
```

(continued)

(bingo.m, *continued*)

```
                    if X(j,k)==n
                        X(j,k)=0;
                    else
                    end
                    if Y(j,k)==n
                        Y(j,k)=0;
                    else
                    end
                end
            end
            if X(1,1)+X(1,2)==0
                xbingo=1;
            elseif X(2,1)+X(2,2)==0
                xbingo=1;
            end
            if Y(1,1)+Y(1,2)==0
                ybingo=1;
            elseif Y(2,1)+Y(2,2)==0
                ybingo=1;
            end
            if xbingo==1&&ybingo==1
                ties=ties+1;keepplaying=0;
            elseif xbingo==1&&ybingo==0
                x=x+1;keepplaying=0;
            elseif xbingo==0&&ybingo==1
                y=y+1;keepplaying=0;
            end
            i=i+1;
        end
    end
    x,y,ties
```

Is Dreidel Fair?

9.1 THE PROBLEM

The ancient Jewish game of dreidel, dating back to at least the early Middle Ages (and perhaps to the time of Christ), is a game of chance played during Hanukkah by two or more players (usually children) with a four-sided top. Each side of the top is marked with one of the four Hebrew letters Nun, Gimel, Hay, and Shin, which we'll call N, G, H, and S, respectively. At the start of a game each of $p \geq 2$ players contributes one unit of money (let's say a dollar) to form an initial pot. (A more traditional pot might consist of pieces of candy, but here I'll make the reward money.) Then, in some sequence agreed upon, the players successively spin the top, which, when it eventually falls over, shows one of its four sides.

The player who has just spun the top receives the following, depending on the letter that shows:

N: nothing,
H: half of the pot,
G: all of the pot,
S: −1, that is, the player must *add* a dollar to the existing pot.

We assume that the top is fair, and so each of these four possible outcomes occurs with probability 1/4 with each spin. The game continues until one of the players spins the first G.

Our question is obvious: the top is fair, yes, but is this a fair game? That is, do all of the players have the same average winnings after they have played a large number of games?

9.2   COMPUTER SIMULATION

Our question can be studied theoretically, but it's not a trivial analysis; it is far more easily and quickly attacked by computer simulation. So let's write a Monte Carlo MATLAB® code that plays ten million games of dreidel for each of the cases of $p = 2$, 3, 4, and 5 players, all the while keeping track of who wins how much. One assumption I'll make in the code **top.m** is that money is infinitely divisible, and so the pot can be halved without limit each time an H shows on the top. Here's how **top.m** works.

The code begins by asking for the value of p (the number of players). Then the p-element vector *winnings* is initialized to zero; *winnings(i)* will always be equal to the present total amount (as the code plays game after game) that player i has won. Then the first of ten million games is started, beginning with the size of *pot* set equal to p (each player contributes a dollar to the pot) and the initial value of *player* set to 1. The variable *keepplaying* is set equal to 1; it is this value that keeps the code in a *while* loop that continues the spinning of the top until the current game ends.

The spinning of our fair top is done by giving the variable *spin* a random value, uniform from 0 to 1. With probability 1/4 we imagine the letter N shows (so "nothing" happens), and we simply pass the top on to the next player (either to *player* + 1 if *player* < p or to player 1 if *player* = p). With probability 1/4 we imagine the letter H shows (and so the pot is halved and the winnings of *player* is incremented by the value of *pot*, which, after having just been halved, is itself half of its value when H showed). Then, as before, we pass the top on to the next player. With probability 1/4 we imagine the letter G shows (and so the current player gets the entire pot), and the game ends, that is, *keepplaying* is set to 0, which will terminate the *while* loop. And finally, with probability 1/4 we imagine the letter S shows (the current player has his winnings *decreased* by one dollar and the pot *increased* by one dollar), and the top is then passed to the next player.

When the *while* loop is exited because *keepplaying* was set to 0, the next game is started, with *pot*, *player*, and *keepplaying* all set back to their starting values. When all ten million games have been played, the results are displayed by the last line of the code, as shown in the following table of average winnings. Clearly, dreidel is not a fair game, which is a curious feature for a game played on a religious holiday! The more players there are, the more unfair it is for the players who wait the longest for their first turn with the top.

Players/Player	1	2	3	4	5
2	1.143	0.857			
3	1.361	0.956	0.68		
4	1.617	1.102	0.757	0.524	
5	1.9	1.267	0.855	0.58	0.398

top.m
```
p=input('How many players?')
winnings=zeros(1,p);
for game=1:10000000
    pot=p;
    keepplaying=1;player=1;
    while keepplaying==1;
        spin=rand;
        if spin<0.25
        elseif spin<0.5
            pot=pot/2;
            winnings(player)=winnings(player)+pot;
        elseif spin<0.75
            winnings(player)=winnings(player)+pot;
            keepplaying=0;
        else
            pot=pot+1;
            winnings(player)=winnings(player)-1;
        end
        if player==p
            player=1;
        else
```

(continued)

(top.m, *continued*)

```
            player=player+1;
        end
      end
    end
    winnings/10000000
```

Hollywood Thrills

10.1 THE PROBLEM

Shortly after his death in Biloxi, Mississippi, in a tragic bungee-jumping accident, the following preliminary movie workup was found in the papers of famed Hollywood director Irving Nutso. Nutso, who as a teenager flunked out of Caltech because he spent all his time watching movies at virtually every theater in the Pasadena, Cucamonga, and Azusa areas around the institute instead of attending classes, was at the time of his death planning his comeback film. With the working title of *Revenge of the Ducks*, it would have been his return to the silver screen after being released from federal prison, having served ten years for alleged involvement with foreign gangsters (who were never charged) in a break-in at the Beverley Hills branch of the International Bank of the Stars.

The millions stolen during that robbery were never recovered, and authorities suspected that Nutso had secretly hidden a lot of cash without telling his gangster friends where. But nothing was ever proved. Still, it might explain where Nutso got funding for *Revenge* (as well as why a 500-foot bungee cord was "accidently" substituted for the correct one when Nutso jumped from a 400-foot-high bridge). While it remains unclear if the film will ever be made, Nutso's preliminary workup does present us with an interesting math problem.

"Revenge of the Ducks" (preliminary)
Irving Nutso

A terrible, never-before-seen bacterial infection has suddenly appeared, worldwide. It seems to have initially appeared in unsanitary theater popcorn butter dispensers, but now it's everywhere. No one is safe, certainly not in airport pay-phone booths or even on Capitol Hill. (*Note to Special Effects:* Can we show some shots of Congresspeople running screaming from their offices using stock footage, or will we have to actually hire some senators?) Once people begin to exhibit symptoms, two-thirds die a terrible, drawn-out death, the specifics of which are so horrible as to be *almost* beyond imagination (but not completely, as close-up details can be graphically depicted in a pre-release movie trailer). Then, just when all hope seems lost, not one but *two* new experimental drugs are announced, each by a different medical team. (All the doctors on each team are beautiful and/or handsome, with two exceptions, as well as all being under the age of 25, even though every one of them has an MD *and* a PhD—the exceptions are that each team will include a goofy but charmingly funny, wise-cracking computer whiz).

Unfortunately each team's preliminary, very uncertain results are based only on animal studies, specifically from lab ducks, as the duck is the only nonhuman animal in which the popcorn butter–loving bacteria thrive. So each drug is separately given to a different group of infected humans. It's impossible to give both drugs to the same person as, together, the drugs interact to form a compound that dissolves bone tissue, a scenario we can gruesomely illustrate in a movie trailer by showing vast flocks of lab ducks collapsing into hysterically quacking blobs as their skeletons turn to mush. This should be quite impressive in 3-D.

The duck data must therefore be supplemented by risky human tests, and so a small number of brave, infected people step forward to serve as volunteers for two fast-result, emergency tests. When drug no. 1 is given to a test group of five infected people, everybody recovers. When drug no. 2 is given to a test group of 12 infected people, however, three die. The president of the United States is then faced with the monster decision of which drug to launch into massive production. Time and money are available only for one or the other. What to do?

The hero of the movie will be portrayed as a brilliant (we'll give him *three* PhDs!) academic mathematician on a sabbatical leave from DUCKSTEW (the Distinguished University for Combining Knowledge with Scientific Technology for the Entire World), who was recently appointed to be special math assistant to the president; it is his enormous task to advise the president on which drug to choose. (I think this role should be played by a well-known teenage heart-throb, who will help attract a younger audience to the film.)

Nutso's notes end here, with just a final scribbled note at the bottom in pencil, saying, "More later, as soon as I get back from my big jump!" We know how *that* ended. How sad.

Anyway, here's the puzzle for you: if *you* were on sabbatical from DUCKSTEW as the president's math adviser, which of the two drugs would you support, the one that had no deaths in its trials or the one that had three deaths? The fate of the world—and the academic reputation of DUCKSTEW—hang in the balance! Only a Hollywood genius like Irv Nutso would have dared to make a gutsy movie like this!

10.2   THEORETICAL ANALYSIS

Most people would be emotionally inclined to go with drug no. 1, which had zero deaths in its test group. That's a perfect record! But, as a DUCKSTEW math prof, you know better: it's actually drug no. 2, with three deaths in its test group, that should be chosen. Here's why. Let's suppose that each of the drugs is actually worthless. That would mean the outcome for each of the test groups was due only to random chance. What we'll do, then, is first calculate the probability that, for each test group, what actually occurred was due only to chance. Then we'll ask ourselves the following question: Which do we believe, "random chance is behind what happened" or "the hypothesis that the drug was worthless has such a small probability—the probability we just computed—that it is more reasonable to reject that hypothesis in favor of believing that the drug *did* have value"? Our choice, then, will be the drug whose trials had the *smaller* probability of occurring only by chance. That is, we'll choose the drug that has the greater likelihood of supporting the rejection of the "worthless" hypothesis.

Let q be the probability an infected person dies after receiving a worthless drug. That means $q = 2/3$. If out of five people all survive (that is, there are zero deaths), then we have an event with probability

$$\binom{5}{0}\left(\frac{2}{3}\right)^0\left(\frac{1}{3}\right)^5 = \frac{1}{3^5} = 0.0041.$$

If out of 12 people at most three die, then we have an event with probability

$$\binom{12}{0}\left(\frac{2}{3}\right)^0\left(\frac{1}{3}\right)^{12} + \binom{12}{1}\left(\frac{2}{3}\right)^1\left(\frac{1}{3}\right)^{11} + \binom{12}{2}\left(\frac{2}{3}\right)^2\left(\frac{1}{3}\right)^{10} +$$
$$\binom{12}{3}\left(\frac{2}{3}\right)^3\left(\frac{1}{3}\right)^9 = \frac{1}{3^{12}} + \frac{24}{3^{12}} + \frac{264}{3^{12}} + \frac{1,760}{3^{12}} = \frac{2,049}{3^{12}} = 0.0038.$$

It is a close call, but even with three deaths during its trials, drug no. 2 offers a slightly better chance of being the more effective drug than does drug no. 1. It's a nonintuitive conclusion, yes, but there it is, a nice illustration of the power of mathematical analysis.

See you at the movies, but probably not at *Revenge of the Ducks*. To be on the safe side, however, just remember to skip putting any butter on your popcorn! And say a prayer for Irv.

The Problem of the n-Liars

11.1 THE PROBLEM

Imagine that we have n people, each of whom tells the truth with probability p when making a statement. That is, if $p=1$ a person never lies (always tells the truth) and if $p=0$ a person always lies. Each of the n people either lies or tells the truth independent of the others. Now, suppose we line up these n people in a row, shoulder to shoulder, from left to right. Starting at the leftmost person, you whisper either "yes" or "no" into his ear, and then that person turns his head and whispers what he heard into the ear of the person next to him. That is, he repeats what he heard *unless he lies* and so whispers "yes" when he actually heard "no," or whispers "no" when he actually heard "yes." This process continues until the rightmost person hears his neighbor whisper one of the two words into his ear; he then speaks out loud either "yes" or "no". What is the probability $Q(n)$ that the word you originally whispered is the word spoken?

Evaluate your answer for the case of $n=41$, with $p=0.99$. What happens as $n \to \infty$?

11.2 THEORETICAL ANALYSIS

In the introduction (I.8), we calculated the probability $P(n)$ of getting an even number of heads in n flips of a coin, with p as the probability of heads on each independent flip:

$$P(n) = \frac{1}{2} + \frac{1}{2}(1-2p)^n.$$

When we did the coin-flipping analysis, it probably seemed to be essentially just an amusing exercise. Now you can see it is actually a doubly amusing exercise! Here's how that earlier analysis applies to this new question.

For a yes to emerge at the end of the chain of liars when it was started with a yes (or for a no to emerge when no started the chain), an even number of lies must have occurred (lies cancel in pairs). This is equivalent to an even number of tails in the coin-flipping problem as each (tails and lies) occurs with probability $1 - p$. Now, if n is even, then getting an even number of tails is the same as getting an even number of heads, and so $Q(n) = P(n)$. That is,

$$Q(n) = \frac{1}{2} + \frac{1}{2}(1-2p)^n, \; n \text{ even.}$$

But if n is odd, then getting an even number of tails is the same as getting an odd number of heads, which occurs with probability $1 - P(n)$. So

$$Q(n) = 1 - \frac{1}{2} - \frac{1}{2}(1-2p)^n = \frac{1}{2} - \frac{1}{2}(1-2p)^n, \; n \text{ odd.}$$

We can write these two $Q(n)$ expressions as a single one, valid for even *and* odd n, as

$$Q(n) = \frac{1}{2} + (-1)^n \frac{1}{2}(1-2p)^n$$

since $(-1)^n = +1$ or -1 for n even or odd, respectively. Thus, we at last have, for any n,

$$Q(n) = \frac{1}{2} + \frac{1}{2}(2p-1)^n.$$

For $p = 0.99$ and $n = 41$ liars, this probability is 0.718393. Also,

$$\lim_{n \to \infty} Q(n) = \frac{1}{2}$$

for all $0 < p < 1$. As the number of liars increases, which word finally emerges at the end of the process is a toss-up for any value of p other than $p = 0$ or $p = 1$. If $p = 1$ (everybody always tells the truth), then obviously $Q(n) = 1$, while if $p = 0$ (everybody always lies), then $Q(n) = 1$ or 0, depending on whether n is even or odd, respectively.

11.3   COMPUTER SIMULATION

The idea behind a computer simulation of the chain of n liars is very straightforward. The code simply counts the number of lies from start to finish, and after the entire chain has been simulated, this count is checked for "evenness." If it is even, then the word that emerges at the end of the chain is the word that started the chain. The brief code **liar.m** does the job, simulating ten million chains of liars, and when run for $n = 41$ and $p = 0.99$ its estimate was $Q(41) = 0.7184214$, impressively close to the theoretical value.

---

liar.m

```
n=41;p=0.99;total=0;
for loop=1:10000000
    lies=0;
    for k=1:n
        if rand>p
            lies=lies+1;
        end
    end
    if lies==2*floor(lies/2)
        total=total+1;
    end
end
total/10000000
```

The Inconvenience of a Law

12.1 THE PROBLEM

In this problem we return to the undocumented resident issue I sketched in the opening of the preface. As you'll recall, there we developed the following mathematical situation, starting with an urn initially containing b black balls and r red balls. We randomly draw one ball at a time; if it's black we put it back in the urn, and if it's red we discard it. We keep doing this until we've removed a fraction f of the red balls. We are interested in how many times, on average, each black ball will be drawn before the process terminates. In particular, we are interested in this average number for the cases of $f = 0.5$, 0.9, and 0.99. Whenever it helps, we'll use the fact that in America, b is in the hundreds of millions and, while r is smaller, it is still in the millions, too (as I write in December 2011, b and r are about 312 million legal residents and 10 to 15 million undocumented residents, respectively).

12.2 THEORETICAL ANALYSIS

To start, let me remind you of a result we derived in Problem 4, concerning the average number μ of attempts until the first S in a random sequence of S's and F's, where p is the probability of an S: $\mu = 1/p$. If we consider drawing a red ball to be an S, then when we start our ball-drawing process we have

$$p = \frac{r}{r+b},$$

and so the average number of drawings until we get the first red ball is

$$\frac{1}{p} = \frac{r+b}{r} = 1 + \frac{b}{r}.$$

Once that first red ball has been drawn (and discarded), we then have an urn with b black balls and $r - 1$ red balls, and so the average number of additional drawings until we get the second red ball is $1 + b/(r - 1)$. We can continue to argue this way until we have reduced the number of red balls to $(1 - f)r$; to get that final red ball will, on average, require $1 + b/(r - rf)$ more drawings. (Since both r and b are large numbers we are not going to worry too much about the slight error made in using rf in place of $rf + n$, where n is a small integer.)

The total number, T, of drawings we make, on average, in reducing the number of red balls from r to $r - rf$ is

$$T = \left(1 + \frac{b}{r}\right) + \left(1 + \frac{b}{r-1}\right) + \left(1 + \frac{b}{r-2}\right) + \ldots + \left(1 + \frac{b}{r-rf}\right),$$

where the number of terms in the sum is $\approx rf$. Thus,

$$T \approx rf + b\sum_{k=0}^{rf}\left(\frac{1}{r-k}\right).$$

The average number of times each black ball will be drawn during this reduction process is T/b, and so a formal answer to our question is

$$\frac{T}{b} \approx \frac{r}{b}f + \sum_{k=0}^{rf}\left(\frac{1}{r-k}\right) \approx \sum_{k=0}^{rf}\left(\frac{1}{r-k}\right),$$

where I've dropped the $(r/b)f$ term, since $(r/b)f \ll 1$.

This is a correct answer, but most people would also find it unsatisfying, too; at the least, it represents a lot of numerical work because the partial sums of the harmonic series increase very slowly. Fortunately, with just a bit more work we can get it into a much nicer form.

To a very good approximation, it has been known since the Swiss-born mathematics genius Euler wrote it (in 1781) that, for $n \geq 1$,

$$\sum_{k=1}^{n}\left(\frac{1}{k}\right) \approx \ln(n) + \frac{1}{2n} + \gamma,$$

where $\gamma = 0.577215664\ldots$ is *Euler's constant*. Even for the very first case of $n = 1$, the error made by the expression on the right is less than 8%, and it gets even smaller as n increases. Now,

$$\sum_{k=0}^{rf}\left(\frac{1}{r-k}\right) = \sum_{k=r-rf}^{r}\left(\frac{1}{k}\right) = \sum_{k=1}^{r}\left(\frac{1}{k}\right) - \sum_{k=1}^{r-rf}\left(\frac{1}{k}\right),$$

where the first equality follows from the observation that the first two sums on the left are the same, term by term, with the second sum simply the first one in reverse order. So, using Euler's approximation on the last two sums on the right, we have

$$\sum_{k=0}^{rf}\left(\frac{1}{r-k}\right) \approx \left[\ln(r) + \frac{1}{2r} + \gamma\right] - \left[\ln(r-rf) + \frac{1}{2(r-rf)} + \gamma\right].$$

Since $1/2r \ll 1$, as is $1/2(r-rf)$, then

$$\sum_{k=0}^{rf}\left(\frac{1}{r-k}\right) \approx \ln(r) - \ln(r-rf) = \ln\left(\frac{r}{r-rf}\right) = \ln\left(\frac{1}{1-f}\right) = -\ln(1-f).$$

This form for the average number of times each black ball is drawn during the process of reducing the number of red balls from r to $r - rf$ is easy to calculate! Notice carefully that there is no dependency on the values of either b or r. This is, I think, totally unexpected. For the values of $f = 0.5, 0.9,$ and 0.99, the average number of times each black ball is drawn is 0.69, 2.3, and 4.6, respectively. Whether or not any of these numbers is "acceptable" in the context of police stopping people at random to find undocumented residents is not a question that can be answered by our analysis. Math gives us the numbers, but people have to make the decision.

A Puzzle for When the Super Bowl

Is a Blowout

13.1 THE PROBLEM

Suppose you and some friends are watching a Super Bowl game on TV and one of the teams is ahead by six touchdowns at the end of the third quarter. Instead of turning to the Weather Channel to get some excitement, do this: challenge your buddies with the following puzzle, and watch their dull, bored eyes light up at the anticipation of working on a good math problem! Be sure to have plenty of extra beer and chips handy because (as is well known) doing real math builds a superappetite.

Let's imagine that the 32 teams in the NFL decide to play a season in the following way. Each team will play every other team once (there is no AFC or NFC, or any other sort of divisional nonsense to muck things up). Further, imagine that all the teams are perfectly matched, and so every game is a toss-up; that is, each team has probability 1/2 of winning any given game. All the games allow for overtime play, and so there are no ties; every game is a win for one team and a loss for another. This will be a pretty long season, with a total of $(31)(32)/2 = 496$ games, but real football fans everywhere will love it! (So will doctors, when players arrive at their offices and hospitals in ambulances at the end of the season, or perhaps even sooner.)

Here's the question: what's the probability that all 32 teams finish the season with *different* records, that is, each team will have a *different* number of games won?

13.2 THEORETICAL ANALYSIS

Let's solve this problem for the general case of n teams and then, when we're finished, we will substitute $n = 32$. There are two key observations that unlock the solution to this problem. The first is that the most games a team can win is $n - 1$ (the number of teams each team plays) because no team can play itself, and the fewest games a team can win is zero. The number of numbers from 0 to $n - 1$ is n, and therefore, when we assign *different* win totals to each team there are just enough to go around. There is a total of $n!$ ways to assign n numbers to n teams. The second key observation is that the probability for each of the $n!$ ways is the same. This follows by symmetry; there is nothing to distinguish one equally skilled team from another. So, if we can calculate the probability for any *one* of the $n!$ ways—let's call that probability P—then the answer to our problem will be just $n!P$.

Suppose we label the teams as no. 1 to no. n, and agree to write the end-of-season games-won totals for the teams from left to right, starting with team no. 1. Then, let's focus our attention on the particular sequence of

$$n - 1, n - 2, n - 3, \ldots, 3, 2, 1, 0.$$

That is, team no. 1 won all of its $n - 1$ games, team no. 2 won all but one of its $n - 1$ games, team no. 3 won all but two of its $n - 1$ games, and so on, all the way to team no. n, which won none of its $n - 1$ games. This means team no. 2 suffered its one loss at the hands of team no. 1 (it couldn't have lost that one game to any other team because then it would have beaten team no. 1, which we know it didn't). In the same way, team no. 3 suffered its two losses at the hands of team no. 1 and team no. 2. And so on. That is, as we move from left to right through the above sequence, each team's losses were at the hands of the teams to its left.

So, the probability of the above particular sequence is the probability team no. 1 won $n - 1$ games times the probability team no. 2 won $n - 2$ games (that team's one loss has already been accounted for in the previous step) times the probability team no. 3 won $n - 3$ games *times* . . . and so on, times the probability team n won zero games. That is,

$$P = \left(\frac{1}{2}\right)^{n-1}\left(\frac{1}{2}\right)^{n-2}\left(\frac{1}{2}\right)^{n-3}\cdots\left(\frac{1}{2}\right)^{0} = \left(\frac{1}{2}\right)^{(n-1)+(n-2)+(n-3)+\cdots+0}$$
$$= \left(\frac{1}{2}\right)^{n(n-1)/2}.$$

The answer to our question, then, is $n!/2^{\frac{n(n-1)}{2}}$. For the NFL, with $n = 32$, this probability is $32!/2^{496} = 1.3 \times 10^{-114}$. Not absolutely impossible, but pretty darn close to impossible.

Darts and Ballistic Missiles

14.1 THE PROBLEM

Two darts players, A and B, stand in front of a square dartboard with dimensions $2R$ by $2R$. On the board is painted a circular target area of radius R with its center at the center of the board. We imagine this center point is the origin of an x,y coordinate system with axes parallel to the edges of the board. A and B each throw a dart at the board, and although each dart does indeed land somewhere on the board, the players have distinctly different throwing techniques. When A throws her dart it lands at a point with coordinates (X, Y), where X and Y are independent random variables, each uniformly distributed from $-R$ to R. When B throws his dart it lands at a point with coordinates (X, X), where X is, as before, a random variable uniformly distributed from $-R$ to R.

As a warm-up for the real puzzler in this problem (soon to come), calculate the probability that A's dart lands in the circular target area (call the result P_A), and then do the same for B's dart (call that probability P_B). These are not difficult calculations to do, and you should be able to verify that $P_A > P_B$ independent of R.

Once you've done that, here's the real problem for you. Instead of X and Y being uniformly distributed, take them as still identically distributed and independent but now *normally* distributed with zero means. That is, their probability density functions are

$$f_X(x) = \frac{1}{\sigma\sqrt{2\pi}} e^{-x^2/2\sigma^2}, \quad -\infty < x < \infty$$

$$f_Y(y) = \frac{1}{\sigma\sqrt{2\pi}} e^{-y^2/2\sigma^2}, \quad -\infty < y < \infty,$$

where σ is a positive parameter (with units of distance) called the *standard deviation*. These are the equations for the famous bell-shaped Gaussian probability curves named after the great German mathematician, Carl Friedrich Gauss (1777–1855).

Now, of course, the dartboard is to be imagined as arbitrarily large, as either dart could potentially land arbitrarily far from the origin in the infinite x,y plane of the board. But our questions remain as before: what are P_A and P_B, the probabilities, respectively, that the darts of A and B land within distance R of the origin of the coordinate system? These new questions are, admittedly, probably now better associated with the impact points of ballistic missiles on distant targets (σ is then a measure of the *miss distance* of the missile) rather than with a darts game, hence the second part of this puzzler's name.

Is it still true that $P_A > P_B$ for all R, as it is in the first case of darts?

14.2 THEORETICAL ANALYSIS

First, the warm-up exercises. When A throws her dart it can land anywhere on the board, with each tiny patch of area as likely to receive the dart as any other tiny patch of area of the same size. So the probability of the dart landing inside the circular target area is simply the ratio of the circular target area to the area of the board. That is,

$$P_A = \frac{\pi R^2}{4R^2} = \frac{\pi}{4} = 0.785.$$

When B throws his dart it lands (by the Pythagorean theorem) a radial distance of $\sqrt{2X^2}$ from the origin. To be in the circular target area, this must be less than or equal to R. Thus,

$$P_B = \text{Prob}(2X^2 \le R) = \text{Prob}\left(-\frac{R}{\sqrt{2}} \le X \le \frac{R}{\sqrt{2}}\right),$$

or, since X is uniform from $-R$ to R,

$$P_B = \frac{2\frac{R}{\sqrt{2}}}{2R} = \frac{1}{\sqrt{2}} = \frac{\sqrt{2}}{2} = 0.707.$$

These two results have no dependency on R, and so A always has the greater probability hitting the circular target area. Let's now see if this remains true when we go to the case when X and Y are normally distributed.

For A we have

$$P_A = \text{Prob}\left(\sqrt{X^2 + Y^2} \leq R\right).$$

Since X and Y are independent, we know their joint probability density function is

$$f_{X,Y}(x,y) = f_X(x)f_Y(y) = \frac{1}{2\pi\sigma^2}e^{-1/2\sigma^2(x^2+y^2)}, \quad -\infty < x, y < \infty,$$

and so, integrating over a circle of radius R centered on the origin of coordinates,

$$P_A = \iint f_{X,Y}(x,y)\,dxdy = \iint f_{X,Y}(x,y)\,dA,$$

where $dA = dxdy$ is the differential area in the x,y coordinate system. Using the well-known trick of converting to polar coordinates, we have $x^2 + y^2 = r^2$ and the differential area is now $dA = rdr\,d\theta$. To cover the circular area, r and θ vary from 0 to R and from 0 to 2π, respectively. So,

$$P_A = \int_0^{2\pi}\int_0^R \frac{1}{2\pi\sigma^2}e^{-r^2/2\sigma^2}rdr\,d\theta = \frac{1}{\sigma^2}\int_0^R r\,e^{-r^2/2\sigma^2}dr.$$

This last integral can be done (and we eventually will), but as you'll soon see, we don't have to do it quite yet.

For B we have

$$P_B = \text{Prob}\left(2X^2 \leq R^2\right) = \text{Prob}\left(-\frac{R}{\sqrt{2}} \leq X \leq \frac{R}{\sqrt{2}}\right)$$

$$= \int_{-R/\sqrt{2}}^{R/\sqrt{2}} \frac{1}{\sigma\sqrt{2\pi}}e^{-x^2/2\sigma^2}\,dx = \frac{2}{\sigma\sqrt{2\pi}}\int_0^{R/\sqrt{2}} e^{-x^2/2\sigma^2}\,dx$$

$$= \frac{1}{\sigma}\sqrt{\frac{2}{\pi}}\int_0^{R/\sqrt{2}} e^{-x^2/2\sigma^2}\,dx,$$

where I've taken advantage of the symmetry of the integrand around $x = 0$. If we change the dummy variable of integration from x to r, with the transformation $r = x\sqrt{2}$, then

$$P_B = \frac{1}{\sigma}\sqrt{\frac{2}{\pi}} \int_0^R e^{-r^2/4\sigma^2} \frac{dr}{\sqrt{2}} = \frac{1}{\sigma\sqrt{\pi}} \int_0^R e^{-r^2/4\sigma^2} \, dr.$$

Now, let's define the function $f(R)$ to be the difference between P_B and P_A. That is,

$$f(R) = P_B - P_A = \frac{1}{\sigma\sqrt{\pi}} \int_0^R e^{-r^2/4\sigma^2} \, dr - \frac{1}{\sigma^2} \int_0^R r e^{-r^2/2\sigma^2} \, dr.$$

Next, recall how to differentiate an integral (for a freshman calculus derivation, see my book *The Science of Radio* [Springer 2001], pp. 415–418):

$$\frac{d}{dR} \int_{\phi_1(R)}^{\phi_2(R)} F(r, R) \, dr = \int_{\phi_1(R)}^{\phi_2(R)} \frac{\partial F}{\partial R} \, dr + F\{\phi_2, R\}\frac{d\phi_2}{dR} - F\{\phi_1, R\}\frac{d\phi_1}{dR}.$$

Since the integrals on the right-hand side in the equation for $f(R)$ have integrands with no dependency on R, we can write df/dR by inspection:

$$\frac{df}{dR} = \frac{1}{\sigma\sqrt{\pi}} e^{-R^2/4\sigma^2} - \frac{1}{\sigma^2} R \, e^{-R^2/2\sigma^2} = \frac{1}{\sigma} e^{-R^2/4\sigma^2}\left[\frac{1}{\sqrt{\pi}} - \frac{1}{\sigma} R \, e^{-R^2/4\sigma^2}\right]$$

$$= \frac{1}{\sigma} e^{-R^2/4\sigma^2} g(R),$$

where

$$g(R) = \frac{1}{\sqrt{\pi}} - \frac{1}{\sigma} R \, e^{-R^2/4\sigma^2}.$$

For any $\sigma > 0$ it is easy to show that $g(R) = 0$ has two positive solutions (I'll prove this in a technical note at the end of this analysis), that is, there are two values of R (call them R1 and R2) such that $g(R1) = g(R2) = 0$, which means that at those values of R we have $df/dR = 0$. Here's why this is useful to know. When $R = 0$ we see that $f = 0$ because both integrals in the definition of f are zero when $R = 0$, and that $df/dR = 1/\sigma\sqrt{\pi} > 0$ when

$R=0$. So, for $0 \leq R < R1$, the function f increases and then, at $R = R1$, f reaches a maximum, and so, for $R > R1$, the function f then decreases until $R = R2$, when f must again reach an extrema, which must now be a minimum. For $R > R2$, f then again increases. Since we have

$$\lim_{R \to \infty} f(R) = \frac{1}{\sigma\sqrt{\pi}} \int_0^\infty e^{-r^2/4\sigma^2}\, dr - \frac{1}{\sigma^2} \int_0^\infty r\, e^{-r^2/2\sigma^2}\, dr,$$

and since the first integral is well known (look in any math tables book, or see my book *Mrs. Perkins's Electric Quilt* [Princeton 2009], pp. 282–283), and the second integral is easy to do, we have

$$\lim_{R \to \infty} f(R) = \frac{1}{\sigma\sqrt{\pi}}(\sigma\sqrt{\pi}) - \frac{1}{\sigma^2}(\sigma^2) = 1 - 1 = 0,$$

and this means, since f is increasing for $R > R2$, that f is negative (that's the only way an increasing function can approach zero).

To summarize what we've learned: starting at $R=0$, we see that $f(R) = P_B - P_A$ increases from zero until $R = R1$, then in the interval $R1 < R < R2$ we see that f decreases and *passes through zero to go negative* until it begins to increase again toward zero when $R > R2$. This means there is some value of R between $R1$ and $R2$—call it R_C—such that

$$f(R) = P_B - P_A = \begin{array}{l} > 0, \quad 0 < R < R_C \\ < 0, \quad R_C < R < \infty \end{array}.$$

Therefore, it is not true, as it was in darts, that $P_A > P_B$, *always*. Rather, if $R < R_C$, then B's missile is the more likely to be in the circular target area, while if $R > R_C$, then it is A's missile that is the more likely to be in the circular target area.

To find the actual value of R_C, let's change variables in the $f(R)$ equation using the transformation $r = u\sigma$. Then $dr = \sigma\, du$, and so

$$f = \frac{1}{\sigma\sqrt{\pi}} \int_0^{R/\sigma} e^{-u^2\sigma^2/4\sigma^2} \sigma\, du - \frac{1}{\sigma^2} \int_0^{R/\sigma} u\sigma\, e^{-u^2\sigma^2/2\sigma^2} \sigma\, du,$$

or, if we write $w = R/\sigma$,

$$f(w) = \frac{1}{\sqrt{\pi}} \int_0^w e^{-u^2/4} \, du - \int_0^w u \, e^{-u^2/2} \, du.$$

We wish to find the value of w that gives $f = 0$. This is done fairly easily by using one of any number of algorithms; I used a particularly popular number-cruncher called the *Newton-Raphson method* (see my book *When Least Is Best* [Princeton 2004], pp. 120–123, for its derivation and historical background). It is an iterative procedure that, given an estimate for w (written as w_n), generates the next (and presumably better) estimate of w_{n+1}:

$$w_{n+1} = w_n - \frac{f(w_n)}{f'(w_n)}$$

where $f(w_n)$ is as given above and $f'(w_n)$ is df/dw, which is given (recall our integral differentiation discussion) by

$$f'(w) = \frac{e^{-w^2/4}}{\sqrt{\pi}} - we^{-w^2/2}.$$

The calculation of $f(w)$ requires good integration software, and I used MATLAB®'s powerful *quad* (for *quadrature*) command. Starting with an initial guess of $w_1 = 1$ (see the technical note below for where this guess comes from), the Newton-Raphson method quickly converges to $w = 1.75294$. That is, $R_C = 1.75294\sigma$.

Technical Note

We now have one final task, that of showing $g(R) = 0$ always has two positive solutions. In Figure 14.2.1 you see plots of $1/\sigma \, R \, e^{-R^2/4\sigma^2}$ for three values of σ (0.5, 1, and 2), and you can see that all three curves peak above the constant $1/\sqrt{\pi}$ (as σ increases, the peak shifts to the right) with the result always being two intersections with the horizontal line. This figure is just for illustration purposes, and it is easy to prove that the peak value is greater than $1/\sqrt{\pi}$ for all $\sigma > 0$ and so $1/\sigma \, R \, e^{-R^2/4\sigma^2} = 1/\sqrt{\pi}$ twice, always.

The maximum of $1/\sigma \, R \, e^{-R^2/4\sigma^2}$ occurs when its derivative equals zero. So,

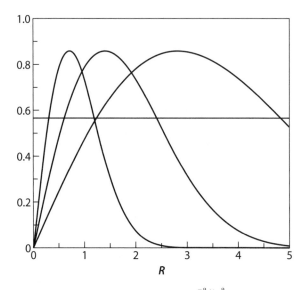

Figure 14.2.1. The maximum of $1/\sigma\, R\, e^{-R^2/4\sigma^2}$ is greater than $1/\sqrt{\pi}$.

$$\frac{1}{\sigma}\left[e^{-R^2/4\sigma^2} + R\left(-\frac{2R}{4\sigma^2}e^{-R^2/4\sigma^2}\right)\right] = \frac{1}{\sigma}e^{-R^2/4\sigma^2}\left[1 - \frac{R^2}{2\sigma^2}\right] = 0$$

when $R = \sigma\sqrt{2}$. This gives a maximum value of

$$\frac{1}{\sigma}\sigma\sqrt{2}\; e^{-2\sigma^2/4\sigma^2} = \sqrt{2}\; e^{-1/2} = \sqrt{\frac{2}{e}},$$

a result independent of σ. Now, $\sqrt{2/e} > 1/\sqrt{\pi}$ because $2/e > 1/\pi$ because $2\pi > e$ because $2\pi > 6$ while $e < 3$.

If you look at the middle curve ($\sigma = 1$, and so $w = R$), you see it crosses the $1/\sqrt{\pi}$ line at about $R \approx 0.5$ and $R \approx 2.5$. These are the values of $R1$ and $R2$, and so R_C is somewhere in between. That's why I picked $w_1 = 1$ (but any initial value in the approximate interval 0.8 to 2.2 would work just as well).

Blood Testing

15.1 THE PROBLEM

Imagine that a very large number of people, N, are to each have their blood tested for a possible dangerous infection. Think, for example, of the millions of men inducted into the armed services during World War II, and the then not-so-uncommon sexually transmitted infection of syphilis. The obvious way to do this is simply to test each person, for a total of N tests. During World War II a different, quite clever alternative approach was actually used, based on the fact that while syphilis wasn't uncommon, most men were actually not infected.

The underlying idea was simple: for some number k, pool the blood samples from k men and test the mixture. If all k men are free of infection, then the test will be negative, and one test clears all k men at once. If the test is positive, however, at least one man is infected, and so then all of them are individually tested, for a total of $k+1$ tests. Taking the probability of being infected as p, this leads to the following two questions: (1) What is the value of k that minimizes the number of expected tests? (2) For that value of k, what is the expected number of tests (and how does it compare with N, the number of tests if each man is individually tested)?

15.2 THEORETICAL ANALYSIS

Since p is the probability a man will test positive, then $1-p$ is the probability he will test negative. Thus, $(1-p)^k$ is the probability a mix of the blood samples from k people will test negative, and $1-(1-p)^k$ is the probability the test will be positive (and so k additional, individual tests will need to be done). So, what we have is

$$1 \text{ test with probability } (1-p)^k$$

and

$$k+1 \text{ tests with probability } 1-(1-p)^k$$

for each blood mixture formed from k people. Since N people will create N/k blood pools, then, if T (a random variable) is the total number of tests required for N people, the average value of T is

$$E(T) = \frac{N}{k}\left\{1 - (1-p)^k + (k+1)\left[1 - (1-p)^k\right]\right\}$$

$$= \frac{N}{k}\left\{(1-p)^k + k + 1 - k(1-p)^k - (1-p)^k\right\}$$

$$= \frac{N}{k}\left\{k + 1 - k(1-p)^k\right\},$$

or,

$$E(T) = N\left\{1 + \frac{1}{k} - (1-p)^k\right\}.$$

To find that k that minimizes $E(T)$, we'll next solve the equation $d\,E(T)/dk = 0$, a calculation that is valid if k is a continuous variable—which of course it isn't. But I'll go ahead with the calculation anyway, as the objection is more a purist's one than anything else. Since

$$(1-p)^k = e^{\ln\{(1-p)^k\}} = e^{k\ln(1-p)},$$

we have

$$\frac{d\,E(T)}{dk} = N\left\{-\frac{1}{k^2} - \ln(1-p)\,e^{k\ln(1-p)}\right\} = 0,$$

or

$$(1-p)^k \ln(1-p) = -\frac{1}{k^2}.$$

For the case of $p \ll 1$ (during the years leading up to World War II a typical value of p for syphilis was something like 0.001 to 0.01), and we can use approximations valid for this situation:

$$(1-p)^k \approx 1 - kp$$

and

$$\ln(1-p) \approx -p.$$

So,

$$-(1-pk)p = -\frac{1}{k^2},$$

or

$$p - p^2 k = \frac{1}{k^2}.$$

Now, since $p \ll 1$, then $p^2 \lll 1$, and so we can drop the $p^2 k$ term. This gives us the answer to our first question:

$$k = \frac{1}{\sqrt{p}}.$$

For example, if $p = 0.01$, then $k = 10$, while if $p = 0.001$, then $k = 32$.

To answer the second question, we'll put $k = 1/\sqrt{p}$ into the expression for $E(T)$ to get

$$E(T) = N\{1 + \sqrt{p} - (1-p)^{1/\sqrt{p}}\},$$

or, using the approximation

$$(1-p)^{1/\sqrt{p}} \approx 1 - p\frac{1}{\sqrt{p}} = 1 - \sqrt{p},$$

we have

$$E(T) = N\{1 + \sqrt{p} - 1 + \sqrt{p}\} = 2N\sqrt{p}.$$

So, if $p = 0.01$, then $E(T) = 0.2N = N/5$, while if $p = 0.001$, then $E(T) = 0.063N = N/16$. These are dramatic reductions in the expected number of tests required.

Big Quotients—Part 2

16.1 THE PROBLEM

A quick recap from the end of Problem 5: we pick N numbers uniformly from 0 to 1, divide the largest by the smallest, and ask for the probability that the result is larger than k, for any $k \geq 1$. This is a significantly more difficult question to answer than was the original Problem 5 question, where we treated in detail the special case of $N = 2$. We could, in theory, proceed just as before, looking now at an N-dimensional unit cube and the subvolumes in that cube that correspond to our problem (in Problem 5 the "cube" was a two-dimensional square and the subvolumes were the shaded triangles in Figure 5.2.1). It's pretty hard to visualize in the Nth dimension, however, so here I'll instead take a purely analytical approach. This will prove to be one of the longer analyses in this book, but I think you'll find it to be a most impressive illustration of the power of mathematics.

16.2 THEORETICAL ANALYSIS

What we wish to calculate is

$$\text{Prob}\left\{\frac{\max\{X_i\}}{\min\{X_i\}} > k\right\}, k \geq 1,$$

where the index i runs from 1 to N and the X_i are independent, uniformly distributed random numbers from 0 to 1. We can reduce this

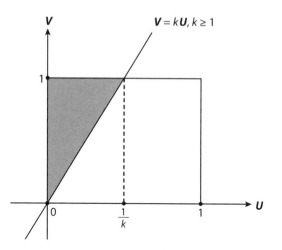

Figure 16.2.1. The probability of the shaded area is $\text{Prob}(V > kU)$.

back down to a two-dimensional problem by defining the random variables $U = \min\{X_i\}$ and $V = \max\{X_i\}$ and asking for

$$\text{Prob}\left\{\frac{V}{U} > k\right\}, k \geq 1.$$

This reduction will come at a price, as you'll soon see. But it will be a price we can pay with just undergraduate mathematics!

In Figure 16.2.1 I've again drawn a two-axis coordinate system, similar to Figure 5.2.1, with V on the vertical axis and U on the horizontal axis. Clearly, both U and V take their values from the interval 0 to 1. The shaded area represents that portion of the unit square (in the first quadrant) where $V \geq kU$, but now the *probability* associated with that area is not the area itself (as it was in Problem 5). That's because neither V nor U is uniformly distributed from 0 to 1, and this is a most important point. To demonstrate the nonuniformity of both V and U is not difficult, and it will be useful to go through such a demonstration as it involves arguments we'll use later in this analysis.

We start by considering the distribution function of V, defined as

$$F_V(v) = \text{Prob}(V \leq v) = \text{Prob}(\max\{X_i\} \leq v).$$

Now, if the largest $X_i \leq v$, then certainly all of the other $X_i \leq v$, too, and so

$$F_V(v) = \text{Prob}(X_1 \leq v, X_2 \leq v, \ldots, X_N \leq v),$$

and so, since the X_i are independent, we then have

$$F_V(v) = \text{Prob}(X_1 \leq v)\,\text{Prob}(X_2 \leq v)\ldots\text{Prob}(X_N \leq v).$$

That is, the distribution function for V is the product of the distribution functions for each of the individual X_i. All of the X_i have the same distribution function because each X_i is *identically* distributed (uniformly random from 0 to 1). Therefore, if we write the distribution function for each X_i as

$$\text{Prob}(X \leq v) = F_X(v),$$

then

$$F_V(v) = F_X^N(v).$$

Okay, but what is $F_X(v)$? Since X is uniform from 0 to 1, then

$$\text{Prob}(X \leq v) = \text{Prob}(0 \leq X \leq v) = v,$$

a linear function. Thus, random variables that are uniform have linear distributions. But we have just shown that

$$F_V(v) = v^N,$$

which is not linear in v. Thus, the random variable $V = \max\{X_i\}$ is not uniformly distributed.

We have to do a slightly more sophisticated analysis to show that $U = \min\{X_i\}$ isn't uniform, either. We start by writing the distribution function of U as

$$F_U(u) = \text{Prob}(U \leq u) = \text{Prob}(\min\{X_i\} \leq u).$$

We have to come up with a new trick now because, unlike what we did in our first argument with the max function, we cannot argue that if the minimum of the $X_i \leq u$, then all the $X_i \leq u$. But what we can write is

$$\text{Prob}\{\min\{X_i\} > u\} = \text{Prob}\{\text{all of the } X_i > u\} = \text{Prob}(X_1 > u) \ldots \text{Prob}(X_N > u)$$

since if the smallest $X_i > u$, then all the other X_i are at least as large. Now,

$$\text{Prob}\{\min\{X_i\} > u\} = 1 - \text{Prob}(\min\{X_i\} \leq u\} = 1 - F_U(u),$$

and so

$$1 - F_U(u) = [1 - \text{Prob}(X_1 \leq u)] [1 - \text{Prob}(X_2 \leq u)] \ldots [1 - \text{Prob}(X_N \leq u)],$$

or

$$1 - F_U(u) = [1 - F_{X_1}(u)] [1 - F_{X_2}(u)] \ldots [1 - F_{X_N}(u)],$$

or, again because each of the X_i has the same distribution function,

$$F_U(u) = 1 - [1 - F_X(u)]^N.$$

Since $F_X(u) = u$, then the distribution of U is

$$F_U(u) = 1 - [1 - u]^N,$$

which is definitely not linear in u. Thus, the random variable $U = \min\{X_i\}$ is not uniformly distributed. This means, as I said before, that we cannot equate the value of the area of the shaded region in Figure 16.2.1 to the value of the probability of the event $V > kU$. What we need to do to calculate that probability is a bit more sophisticated.

The probability density function of V, written as $f_V(v)$, is a function such that

$$\text{Prob}(V \leq v) = \int_{-\infty}^{v} f_V(z)\, dz = F_V(v),$$

where z is simply a dummy variable of integration. That is, the distribution is the integral of the density. In the same way, the probability density function of U is $f_U(u)$, where

$$\text{Prob}(U \le u) = \int_{-\infty}^{u} f_U(z)\, dz = F_U(u).$$

If we differentiate both sides of these two equations, we have

$$f_V(v) = \frac{d}{dv} F_V(v)$$

and

$$f_U(u) = \frac{d}{du} F_U(u).$$

That is, the density of a random variable is the derivative of the distribution.

The densities for the maximum and the minimum functions, in the general case of identically distributed X_i with arbitrary density (not necessarily restricted to being uniform from 0 to 1), are

$$f_V(v) = N F_X^{N-1}(v) f_X(v)$$

and

$$f_U(u) = N[1 - F_X(u)]^{N-1} f_X(u).$$

We'll use the distribution and density functions for both U and V again later in this book (Problem 19), and so these calculations will have a further payoff for us when we are done with our present problem.

When we have a problem involving two random variables, as we do here, we can define what is called the *joint probability density*, written as $f_{U,V}(u, v)$—recall our analysis in Problem 14. Just as integrating the density over an interval gives us the probability a random variable has a value from that interval, integrating a joint density over a region gives us the probability our two random variables have values that locate a point in that region—again, recall Problem 14. So, the probability we are after, Prob($V > kU$), is the integral of the joint density of U and V over the shaded region of Figure 16.2.1. That is,

$$\text{Prob}(V > kU) = \int_0^1 \int_0^{v/k} f_{U,V}(u,v)\,du\,dv,$$

where the inner u-integration is along a *horizontal* strip (distance v up from the U axis), and the outer v-integration moves the horizontal strip upward, starting from $v = 0$ and ending at $v = 1$. That is, our double integral *vertically scans* the shaded region. What we need to do next, then, is find the joint density of U and V so that we can perform the integrations.

If U and V were independent, then the joint density would simply be the product of the individual densities—you'll recall this is what we were able to do in Problem 14. Our U and V in this problem, however, are not independent. After all, if I tell you what U is, then you immediately know V is bigger. Or if I tell you what V is, then you immediately know U is smaller. That is, knowledge of either U or of V gives you knowledge of the other, and so

$$f_{U,V}(u,v) \neq f_U(u) f_V(v).$$

Okay, this is what we can't do; what *can* we do?

Generalizing to two random variables considered together, we can define their joint distribution function $F_{U,V}(u,v)$ as

$$F_{U,V}(u,v) = \text{Prob}(U \leq u, V \leq v) = \int_{-\infty}^u \int_{-\infty}^v f_{U,V}(z_1,z_2)\,dz_1\,dz_2,$$

and so extending the idea of getting a density by differentiating a distribution, the joint density $f_{U,V}(u,v)$ is given by a double differentiation:

$$f_{U,V}(u,v) = \frac{\partial^2}{\partial u \partial v} F_{U,V}(u,v).$$

What we need to do, then, is find the joint distribution of U and V, then partially differentiate it twice to get the joint density, and then integrate that density over the shaded region in Figure 16.2.1. This sounds like a lot to do, but in fact, it really isn't that hard, as I'll now show you.

To find $F_{U,V}(u,v)$, there are two cases to consider, that of $u \geq v$ and $u < v$. For the first case, the upper bound u on the minimum X_i is greater than the upper bound on the maximum X_i, and so of course the value of u really plays no role; after all, if the maximum X_i can be no larger than v, then we know that all the $X_i \leq v$, and so

$$
\begin{aligned}
F_{U,V}(u,v) &= \text{Prob}(X_1 \leq v, X_2 \leq v, \ldots X_N \leq v) \\
&= \text{Prob}(X_1 \leq v)\,\text{Prob}(X_2 \leq v) \ldots \text{Prob}(X_N \leq v) \\
&= F_X^N(v), u \geq v.
\end{aligned}
$$

Since there is no u dependency in $F_{U,V}(u,v)$ for $u \geq v$, then

$$
f_{U,V}(u,v) = 0, u \geq v.
$$

For the second case, where $u < v$, the upper bound on the minimum is less than the upper bound on the maximum. All of the $X_i \leq v$, of course, but now at least one of the $X_i \leq u$, too. What this means is that *all* of the X_i *cannot* be in the interval u to v. That is, for $u < v$,

$$
F_{U,V}(u,v) = \text{Prob}(\textit{all the } X_i \leq v) - \text{Prob}(\textit{all the } X_i \text{ both} > u \text{ and} \leq v).
$$

The first probability on the right is

$$
\text{Prob}(\textit{all the } X_i \leq v) = \text{Prob}(X_1 \leq v)\,\text{Prob}(X_2 \leq v) \ldots \text{Prob}(X_N \leq v) = F_X^N(v),
$$

while the second probability on the right is

$$
\begin{aligned}
\text{Prob}(\textit{all the } X_i \text{ both} > u \text{ and} \leq v) &= \text{Prob}(u < X_1 \leq v) \ldots \text{Prob}(u < X_N \leq v) \\
&= [F_X(v) - F_X(u)]^N.
\end{aligned}
$$

Thus,

$$
F_{U,V}(u,v) = F_X^N(v) - [F_X(v) - F_X(u)]^N, u < v.
$$

Differentiating first with respect to v, we have

$$\frac{\partial}{\partial v} F_{U,V}(u,v) = NF_X^{N-1}(v) f_X(v) - N[F_X(v) - F_X(u)]^{N-1} f_X(v),$$

and then differentiating again with respect to u,

$$\frac{\partial^2}{\partial u \partial v} F_{U,V}(u,v) = f_{U,V}(u,v) = N(N-1)[F_X(v) - F_X(u)]^{N-2} f_X(v) f_X(u), \; u < v,$$

which we see, as claimed earlier, is NOT

$$f_U(u) f_V(u) = \{N[1 - F_X(u)]^{N-1} f_X(u)\}\{NF_X^{N-1}(v) f_X(v)\}.$$

Since all the X_i in our problem are uniform from 0 to 1, we have

$$f_X(v) = \begin{array}{ll} 1, & 0 \le v \le 1 \\ 0, & \text{otherwise} \end{array}$$

$$f_X(u) = \begin{array}{ll} 1, & 0 \le u \le 1 \\ 0, & \text{otherwise} \end{array}$$

$$F_X(v) = \begin{array}{ll} v, & 0 \le v \le 1 \\ 0, & v < 0 \\ 1, & v > 1 \end{array}$$

$$F_X(u) = \begin{array}{ll} u, & 0 \le u \le 1 \\ 0, & u < 0 \\ 1, & u > 1 \end{array}.$$

If you look back at Figure 16.2.1 you'll see that over the entire shaded region of integration we have $u < v$, and so using the above expressions for the distributions and densities involved,

$$f_{U,V}(u,v) = N(N-1)(v-u)^{N-2}.$$

Thus, the formal answer to our question is the probability integral

$$\text{Prob}(V > kU) = N(N-1) \int_0^1 \int_0^{v/k} (v-u)^{N-2} du \, dv.$$

Notice that if $N = 2$ (the problem we solved by geometric means back in Problem 5) this integral reduces to

$$2\int_0^1 \int_0^{v/k} du \, dv = 2\int_0^1 (u)|_0^{v/k} dv = 2\int_0^1 \frac{v}{k} dv = \frac{2}{k}\left(\frac{1}{2}v^2\Big|_0^1\right) = \left(\frac{2}{k}\right)\left(\frac{1}{2}\right) = \frac{1}{k},$$

just as we found before.

16.3   COMPUTER SIMULATION

For $N = 3$, the case we could only *simulate* (with the code **ratio2.m**) at the end of Problem 5, our probability integral becomes

$$6 \int_0^1 \int_0^{v/k} (v - u) \, du \, dv = 6 \int_0^1 \left(vu - \frac{1}{2} u^2 \right) \Big|_0^{v/k} \, dv = 6 \int_0^1 \left(\frac{v^2}{k} - \frac{v^2}{2k^2} \right) dv$$

$$= 6 \left(\frac{v^3}{3k} - \frac{v^3}{6k^2} \right) \Big|_0^1 = \frac{2}{k} - \frac{1}{k^2}.$$

If you plug in the values of k that I used in Table 5.3.2 for the $N = 3$ case, you'll find there is remarkably good agreement between the simulation results from **ratio2.m** and this theoretical result.

You should now be able to evaluate our probability integral for the $N = 4$ case, as well as modify **ratio2.m** to simulate the $N = 4$ case, and so again compare theory with experiment. Try it! If you feel really ambitious, do the integral in general—it actually isn't that difficult—and show that

$$\text{Prob}(V > kU) = 1 - \left(1 - \frac{1}{k} \right)^{N-1}.$$

With this result, we can now answer the following question: how many numbers (all independently and randomly selected uniformly from 0 to 1) must be taken so that, with probability P, the ratio of the largest to the smallest is greater than k? We have

$$P = 1 - \left(1 - \frac{1}{k} \right)^{N-1},$$

or

$$\left(1 - \frac{1}{k} \right)^{N-1} = 1 - P,$$

and so, taking logs on both sides and solving for N, we have

$$N = \frac{\log(1 - P)}{\log\left(1 - \frac{1}{k} \right)} + 1.$$

For example, if $k = 4$ and $P = 0.9$, then

$$N = \frac{\log(0.1)}{\log(0.75)} + 1 = 9.004,$$

and so $N = 10$.

To Test—Or Not to Test?

17.1 THE PROBLEM

In our modern times, new and wonderful advances are seemingly achieved between the last time we turned the TV off and when we turn it back on. This is particularly true in medicine, where apparently there is a treatment for every ill that afflicts the human body, ranging over the entire spectrum from warts and dry eyes to erectile dysfunction and urinary incontinence to asthma and bowel disease to. . . . Well, you get the idea, and the message, too: drug companies have spent billions of dollars telling us to "ask your doctor" about the latest pills, patches, injections, inhalers, and drops to arrive at the local drugstore.

We are also told to have regular diagnostic tests for all sorts of awful things, from mammograms for women (breast cancer) to blood PSA levels for men (prostate cancer) to colonoscopies, blood sugar levels, abdominal ultrasounds, and chest X-rays for everybody (to detect colon cancer, diabetes, aortic aneurysm, and lung cancer, respectively). While any of those tests is almost surely "good for you," such testing does come with a potentially serious issue. It is, alas, simply not perfect. That is, any diagnostic test has associated with it two quite different statistical errors, the false positive and the false negative.

A false positive occurs when a test says you have the condition tested for when you actually don't. The usual result in that case is that you are scared silly (at least at first) and then later (more extensive and usually more expensive) testing reveals you are really okay. A false negative occurs when a test says you don't have the condition tested for when you actually do. This is potentially a far more serious outcome than is a false positive result because it can lull you into a sense of security and encourage you to take no further action—and then, a month or so later, perhaps, you suddenly drop dead. That's particularly sad when you think that a ten-cent-a-day pill might have given you another fifty years!

With these scary considerations in mind, you can understand why it is an important goal in any medical trial of a new diagnostic test to determine the statistical behavior of the test. For example, let's suppose you have decided to take a newly developed test for a terrible condition, one whose very name is unspeakable in other than a hushed whisper: awfulitis. It is estimated that only one in ten thousand people have awfulitis, but since it either kills you with probability 0.99 or, with probability 0.01, makes you *wish* it had killed you, you are plenty worried about it. Now, clearly, either you have awfulitis or you don't, which we'll denote by $\mathbf{A}$ and $\overline{\mathbf{A}}$, respectively. When you take the test it will, equally clearly, say either you have awfulitis or that you don't, which we'll denote by $\mathbf{T}$ and $\overline{\mathbf{T}}$, respectively.

The probability of a false positive when you take the test, in the notation of conditional probability, is written as $P(\mathbf{T}|\overline{\mathbf{A}})$, which is read as "the probability the test says you have awfulitis given that you actually don't." (The "given" part is always to the right of the vertical bar.) The probability of a false negative when you take the test is written as $P(\overline{\mathbf{T}}|\mathbf{A})$, which is read as "the probability the test says you don't have awfulitis given that you actually do."

Extensive trials, using groups of people *known* either to have or not to have awfulitis, have determined that

$$P(\mathbf{T}\,|\,\mathbf{A}) = 0.95$$

and

$$P(\overline{\mathbf{T}}\,|\,\overline{\mathbf{A}}) = 0.95.$$

These two results contain within them the probabilities of both the false negative and false positive errors, because we can write

$$P(\mathbf{T}\,|\,\mathbf{A}) + P(\overline{\mathbf{T}}\,|\,\mathbf{A}) = 1$$

and

$$P(\mathbf{T} \mid \overline{\mathbf{A}}) + P(\overline{\mathbf{T}} \mid \overline{\mathbf{A}}) = 1.$$

In each of these two equations we hold the given part fixed, and then add the conditional probabilities of each of the two possible test outcomes. The sum in each case must be 1, because "given a fixed condition," the test will say *something*. So, we have

$$P(\overline{\mathbf{T}} \mid \mathbf{A}) = 1 - P(\mathbf{T} \mid \mathbf{A}) = 0.05 \text{ (probability of false negative)}$$

and

$$P(\mathbf{T} \mid \overline{\mathbf{A}}) = 1 - P(\overline{\mathbf{T}} \mid \overline{\mathbf{A}}) = 0.05 \text{ (probability of false positive)}.$$

To most people, these seem like pretty impressive numbers. Zero probability for both kinds of errors would be best, of course, but 0.05 looks small. But is 0.05 small enough? The way to answer that question is to ask yourself what you would do after being told the test result. Suppose you are told that the test says you have awfulitis. But do you? Or suppose the test says you don't have awfulitis. Is that true? In the first case, what you'd want to calculate is $P(\mathbf{A} \mid \mathbf{T})$, the conditional probability that you have awfulitis given that the test says you do. And in the second case, you'd want to calculate $P(\overline{\mathbf{A}} \mid \overline{\mathbf{T}})$, or the conditional probability that you don't have awfulitis given that the test says you don't. My personal experience is that many family doctors often do not know how to do these calculations, or even that such calculations can be done. So, what are these conditional probabilities for awfulitis?

17.2 THEORETICAL ANALYSIS

From conditional probability theory we have the fundamental definition

$$P(\mathbf{A} \mid \mathbf{T}) = \frac{P(\mathbf{AT})}{P(\mathbf{T})}.$$

Like most definitions in mathematics this has not simply been pulled out of the air but rather is motivated by a specific, special case. Imagine a finite sample space with N different, *equally likely* sample points. Of these N points, n_A are associated with $\mathbf{A}$, n_T are associated with $\mathbf{T}$, and n_{AT} are common to both $\mathbf{A}$ and $\mathbf{T}$. Then

$$P(\mathbf{T}) = \frac{n_T}{N}, P(\mathbf{AT}) = \frac{n_{AT}}{N}, P(\mathbf{A}\,|\,\mathbf{T}) = \frac{n_{AT}}{n_T},$$

from which we immediately have our definition. In exactly the same way we have

$$P(\mathbf{T}\,|\,\mathbf{A}) = \frac{P(\mathbf{AT})}{P(\mathbf{A})},$$

which says

$$P(\mathbf{AT}) = P(\mathbf{T}\,|\,\mathbf{A})P(\mathbf{A}),$$

and so

$$P(\mathbf{A}\,|\,\mathbf{T}) = \frac{P(\mathbf{T}\,|\,\mathbf{A})P(\mathbf{A})}{P(\mathbf{T})}.$$

The theorem of total probability says[1]

$$P(\mathbf{T}) = P(\mathbf{T}\,|\,\mathbf{A})P(\mathbf{A}) + P(\mathbf{T}\,|\,\overline{\mathbf{A}})P(\overline{\mathbf{A}}),$$

and so

$$P(\mathbf{A}\,|\,\mathbf{T}) = \frac{P(\mathbf{T}\,|\,\mathbf{A})P(\mathbf{A})}{P(\mathbf{T}\,|\,\mathbf{A})P(\mathbf{A}) + P(\mathbf{T}\,|\,\overline{\mathbf{A}})P(\overline{\mathbf{A}})}.$$

We know all of the probabilities on the right (one person in 10,000 having awfulitis means $P(\mathbf{A}) = 0.0001$ and so $P(\overline{\mathbf{A}}) = 0.9999$):

$$P(\mathbf{A}\,|\,\mathbf{T}) = \frac{(0.95)(0.0001)}{(0.95)(0.0001) + (0.05)(0.9999)} = \frac{0.95}{(0.95) + (0.05)(9,999)}$$

$$= \frac{0.95}{(0.95) + (499.95)} = 1.9 \times 10^{-3},$$

a result that, without fail, astonishes everybody. The test is almost surely wrong if it says you have awfulitis!

This is an important calculation to perform because in addition to making you feel better—*Thank God, I'm (almost surely) not going to die!*—the calculation tells researchers what has to change to improve the test. $P(\mathbf{A}|\mathbf{T})$ is small because of the large term in the denominator due to $P(\mathbf{T}|\overline{\mathbf{A}})P(\overline{\mathbf{A}})$. There's nothing we can do about the $P(\overline{\mathbf{A}})$ factor; even if you could do something about it, there isn't much left to do, as it's practically 1 already and, if it were 1 (which means nobody has awfulitis), that would eliminate the need in the first place for the test. Rather, it's the factor $P(\mathbf{T}|\overline{\mathbf{A}}) = 1 - P(\overline{\mathbf{T}}|\overline{\mathbf{A}})$ that needs to be greatly reduced. In other words, $P(\overline{\mathbf{T}}|\overline{\mathbf{A}}) = 0.95$, which initially looked so good, actually isn't nearly good enough, and must be increased. Suppose, for example, that we could make $P(\overline{\mathbf{T}}|\overline{\mathbf{A}}) = 0.9999$. *That* sure looks good! Well, let's see if it is.

This new value for $P(\overline{\mathbf{T}}|\overline{\mathbf{A}})$ gives $P(\mathbf{T}|\overline{\mathbf{A}}) = 0.0001$, and so

$$P(\mathbf{A}|\mathbf{T}) = \frac{(0.95)(0.0001)}{(0.95)(0.0001) + (0.0001(0.9999)} = \frac{0.95}{(0.95) + (0.0001)(9,999)}$$
$$= \frac{0.95}{(0.95) + (0.9999)} = 0.487,$$

which is a big improvement, but the test is still wrong more often than it's right.

Okay, let's stop guessing and approach this in reverse. What must $P(\overline{\mathbf{T}}|\overline{\mathbf{A}})$ be so that $P(\mathbf{A}|\mathbf{T}) = 0.95$? We have

$$P(\mathbf{A}|\mathbf{T}) = \frac{P(\mathbf{T}|\mathbf{A})P(\mathbf{A})}{P(\mathbf{T}|\mathbf{A})P(\mathbf{A}) + [1 - P(\overline{\mathbf{T}}|\overline{\mathbf{A}})]P(\overline{\mathbf{A}})},$$

and so, solving for $P(\overline{\mathbf{T}}|\overline{\mathbf{A}})$ gives

$$P(\overline{\mathbf{T}}|\overline{\mathbf{A}}) = \frac{P(\mathbf{A}|\mathbf{T})P(\mathbf{T}|\mathbf{A})P(\mathbf{A}) + P(\mathbf{A}|\mathbf{T})P(\overline{\mathbf{A}}) - P(\mathbf{T}|\mathbf{A})P(\mathbf{A})}{P(\mathbf{A}|\mathbf{T})P(\overline{\mathbf{A}})}$$
$$= \frac{(0.95)(0.95)(0.0001) + (0.95)(0.9999) - (0.95)(0.0001)}{(0.95)(0.9999)}$$
$$= \frac{0.95 + 9,999 - 1}{9,999} = \frac{9,998.95}{9,999} = 0.999995.$$

That is, given a million people who are known *not* to have awfulitis, the test must say (erroneously) that, at most, five of them have awfulitis.

For our last calculation, that of $P(\overline{\mathbf{A}}\,|\,\overline{\mathbf{T}})$, doing the same sort of analysis as we did for $P(\mathbf{A}|\mathbf{T})$ should allow you to derive

$$P(\overline{\mathbf{A}}\,|\,\overline{\mathbf{T}}) = \frac{P(\overline{\mathbf{T}}\,|\,\overline{\mathbf{A}})P(\overline{\mathbf{A}})}{P(\overline{\mathbf{T}}\,|\,\mathbf{A})P(\mathbf{A}) + P(\overline{\mathbf{T}}\,|\,\overline{\mathbf{A}})P(\overline{\mathbf{A}})} = \frac{(0.95)(0.9999)}{(0.05)(0.0001) + (0.95)(0.9999)}$$
$$= \frac{(0.95)(9,999)}{(0.05) + (0.95)(9,999)} = \frac{9,499.95}{9,499.1} = 0.999995.$$

That is, as it stands the test is a very good test if it declares you to be free of awfulitis. So, the bottom line concerning our diagnostic test is: if the news is bad, then certainly do a follow-up test (and you're probably okay anyway), while if the news is good, you (almost certainly) *are* okay.

Historical Note

This sort of probabilistic analysis (called *Bayesian analysis*, after the English minister Thomas Bayes [1701–1761], who initiated such calculations in a posthumously published 1764 paper in the *Philosophical Transactions of the Royal Society of London*) has been unfairly burdened over the years with some unfortunate applications. The best known of these misuses is probably Laplace's infamous calculation of the probability the sun will rise tomorrow.

Beginning in 1744, the French mathematician Pierre-Simon Laplace (1749–1827) began writing on what is known as his "law of succession," which answers the following question: if an event in a repeatable probabilistic experiment has been observed to happen n times in a row, what is the probability that event will happen yet again in the very next experiment? The experiment might be the flipping of a coin or, in Laplace's erroneous example, the rising of the sun. Laplace's starting point in his analysis was with the assumption of total ignorance, which, in the coin-flipping example, means all possible values for the probability of heads are equally likely (uniform from 0 to 1). Since this analysis has historical value, let me now show you a balls-and-urns derivation of Laplace's result.

Suppose you are confronted with $N+1$ urns, each of which contains N balls. The urns are numbered from 0 to N, with urn k containing k black balls and $N-k$ white balls (urn 0 has all white balls and urn N

has all black balls). Now, imagine that you select an urn at random and then make n consecutive drawings from it, each time replacing the ball after observing its color. If all n of those drawings resulted in a black ball, then what's the probability the next drawing will also result in a black ball? This is obviously a *conditional* probability, in that if $\mathbf{A}$ is the event "n drawings of n black balls" and $\mathbf{B}$ is the event "next drawing will be a black ball," then what we are asking for is $P(\mathbf{B}|\mathbf{A})$. The fact that we first select (at random) one urn from urns representing all possible combinations of black and white balls is the way this problem models Laplace's assumption of total ignorance.

Given that we first select urn k, then the probability of drawing a black ball n straight times (event $\mathbf{A}$) is

$$P(\mathbf{A}\mid \mathbf{U}_k) = \left(\frac{k}{N}\right)^n,$$

where $\mathbf{U}_k$ is the event "urn k selected." From the theorem of total probability we can write

$$P(\mathbf{A}) = \sum_{k=0}^{N} P(\mathbf{A}\mid \mathbf{U}_k)\, P(\mathbf{U}_k),$$

and, since $P(\mathbf{U}_k)=1/(N+1)$, we have

$$P(\mathbf{A}) = \frac{1}{N+1}\sum_{k=0}^{N}\left(\frac{k}{N}\right)^n = \frac{1^n + 2^n + 3^n + \cdots + N^n}{(N+1)\,N^n}.$$

$\mathbf{AB}$ is the joint event of "n drawings of n black balls *and* the next (that is, the $n+1$st drawing) is also a black ball," which means $n+1$ drawings have produced $n+1$ black balls. From our last result, for $P(\mathbf{A})$, we have $P(\mathbf{AB})$ given by just replacing n with $n+1$ to get

$$P(\mathbf{AB}) = \frac{1^{n+1} + 2^{n+1} + 3^{n+1} + \cdots + N^{n+1}}{(N+1)N^{n+1}}.$$

Also, note that $P(\mathbf{AB}) = P(\mathbf{B})$ since if we have had $n+1$ consecutive black balls drawn (event $\mathbf{B}$), then event $\mathbf{A}$ (n consecutive black balls drawn) has also occurred.

Thus, the (conditional) probability of interest, $P(\mathbf{B}|\mathbf{A})$, is

$$P(\mathbf{B} \mid \mathbf{A}) = \frac{P(\mathbf{AB})}{P(\mathbf{A})} = \frac{P(\mathbf{B})}{P(\mathbf{A})} = \frac{1^{n+1} + 2^{n+1} + 3^{n+1} + \cdots + N^{n+1}}{1^n + 2^n + 3^n + \cdots + N^n} \left(\frac{N^n}{N^{n+1}} \right).$$

The sum in the denominator in this expression can be approximated by the integral of x^n as x varies from 0 to N, and we can do the same for the numerator. Thus,

$$P(\mathbf{B} \mid \mathbf{A}) \approx \left(\frac{1}{N} \right) \frac{\int_0^N x^{n+1} dx}{\int_0^N x^n dx} = \left(\frac{1}{N} \right) \frac{\left[1/(n+2) x^{n+2} \right]_0^N}{\left[1/(n+1) x^{n+1} \right]_0^N}$$

$$= \left(\frac{1}{N} \right) \left(\frac{n+1}{n+2} \right) \left(\frac{N^{n+2}}{N^{n+1}} \right) = \frac{n+1}{n+2}.$$

As a colorful illustration of this result, Laplace replaced the successive drawings of black balls out of an urn with the successive risings of the sun. In his 1814 *A Philosophical Essay on Probabilities* he wrote, "Placing the most ancient epoch of history at five thousand years ago, or at 1,826,213 days, and the sun having risen constantly in the interval at each revolution of twenty-four hours, it is a bet of 1826214 to one that it will rise again tomorrow." Analysts have ever since ridiculed Laplace's words as nonsense, as describing a situation that violates the very ideas of a "repeatable probabilistic experiment." There is nothing at all random about the sun rising, with Newton's inverse-square law of gravity and celestial mechanics being behind that daily event.

Was Laplace serious with his example? He was, after all, a genius. In fact, Laplace knew perfectly well the example was flawed, as the very next sentence (one usually not quoted by his critics) after the one quoted above is, "But this number is comparably greater for him who, *recognizing in the totality of phenomena the principal regulator of days and seasons* [my emphasis], sees that nothing at the present moment can arrest the course of it [the rising of the sun]." Laplace, in looking for a dramatic illustration for his law of succession, simply went a step too far, with the unfortunate consequence of throwing a shadow over Bayesian analysis in general. The use we made of it in the medical test analysis earlier in this chapter is perfectly fine.

NOTE

1. This result should be self-evident by writing the right-hand side as

$$\frac{P(\mathbf{AT})}{P(\mathbf{A})}P(\mathbf{A}) + \frac{P(\overline{\mathbf{A}}\mathbf{T})}{P(\overline{\mathbf{A}})}P(\overline{\mathbf{A}}) = P(\mathbf{AT}) + P(\overline{\mathbf{A}}\mathbf{T}),$$

which is the probability **T** occurs and **A** does, too, plus the probability **T** occurs and **A** doesn't. That is, the sum is simply the probability **T** occurs, *and we don't care whether* **A** *does or doesn't.* But that is precisely $P(\mathbf{T})$.

Average Distances on a Square

18.1 THE PROBLEM(S)

This is the longest analysis in the book (it's actually three related problems), designed to show you that computer simulation can definitely have its place—theory is always desirable, of course, but often it is simulation that will save your sanity in the short run when you need an answer fast! Our general question is easy to state: given a unit square, and two points picked at random from that square, what's the average distance between the two points?

Stated this way, however, the question is not well defined, as there are at least three ways we can interpret what the words "two points picked at random from a square" mean. (See the last section of this chapter for yet a fourth interpretation, one that I analyzed in an earlier book.) They are (where the numbers in parentheses are the answers):

- (a) pick one point on a side and the other on an adjacent side (0.7652);
- (b) pick one point on a side and the other on the opposite side (1.0766);
- (c) pick both points from anywhere on the boundary or the interior (0.5214).

Version (c) might occur, for example, in a study of two wild animals coexisting on a common square patch of land. All of these interpretations have two common features: their theoretical solutions are increasingly complicated, with (a) being the easiest but still plenty challenging, while all have very simple simulations.

18.2   THEORETICAL ANALYSES

Here I'll take you through the analyses for (a) and (b), and when we are finished, I think you'll be more than willing to skip the theory for (c), which is much more involved (although I'll sketch its start for you, too). Analytically, this is probably one of the two toughest solved problems we'll tackle in this book (Problem 16 is the other one). Since in (a) we are picking our two points from adjacent sides, we can imagine our unit square positioned as shown in Figure 18.2.1, with one of the points on the y-axis and the other on the x-axis. The distance between the two points is the value of the random variable R, where

$$R = \sqrt{X^2 + Y^2}$$

where X and Y are independent, uniformly distributed (each over 0 to 1) random variables. Obviously, R varies from 0 to $\sqrt{2}$.

Our problem is the calculation of the average (or expectation) of R, given by

$$E(R) = \int_0^{\sqrt{2}} r f_R(r) dr$$

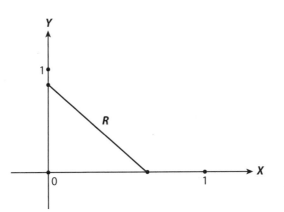

Figure 18.2.1. A random line connecting adjacent sides of a unit square.

where $f_R(r)$ is the *probability density function* of **R**. So, our first task is to calculate $f_R(r)$, and to do that we'll first find the probability *distribution function* $F_R(r) = \text{Prob}(\textbf{R} \leq r) = \int_{-\infty}^{r} f_R(u)\, du$, and then differentiate it to get the density.

We start by writing

$$F_R(r) = \text{Prob}\left(\sqrt{X^2 + Y^2} \leq r\right) = \text{Prob}\left(Y^2 \leq r^2 - X^2\right)$$
$$= \text{Prob}\left(-\sqrt{r^2 - X^2} \leq Y \leq \sqrt{r^2 - X^2}\right).$$

To calculate this last probability we have to consider the two cases of $0 \leq r \leq 1$ and $1 \leq r \leq \sqrt{2}$, as shown in Figures 18.2.2 and 18.2.3. In both cases, since **X** is uniform from 0 to 1, the probability of $Y \leq \sqrt{r^2 - X^2}$ is the area under the curve $y \leq \sqrt{r^2 - x^2}$ *that is also in the unit square* (the italicized words are crucial in the case of $1 \leq r \leq \sqrt{2}$ shown in Figure 18.2.3). The distribution function of **R** is

$$F_R(r) = \frac{\int_0^r \sqrt{r^2 - x^2}\, dx, \ \ 0 \leq r \leq 1}{\sqrt{r^2 - 1} + \int_{\sqrt{r^2 - 1}}^1 \sqrt{r^2 - x^2}\, dx, \ \ 1 \leq r \leq \sqrt{2}.}$$

From integral tables we have

$$\int \sqrt{r^2 - x^2}\, dx = \frac{x\sqrt{r^2 - x^2}}{2} + \frac{r^2}{2} \sin^{-1}\left(\frac{x}{r}\right),$$

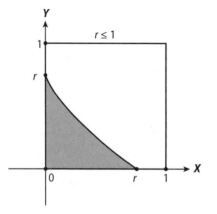

Figure 18.2.2. $0 \leq r \leq 1$.

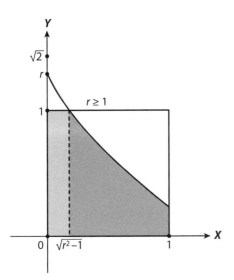

Figure 18.2.3. $1 \le r \le \sqrt{2}$.

and so

$$F_R(r) = \frac{r^2}{2} \sin^{-1}(1) = \frac{\pi r^2}{4}, \quad 0 \le r \le 1.$$

Differentiating then gives us the density function

$$f_R(r) = \frac{\pi}{2}, \quad 0 \le r \le 1.$$

For the case of $1 \le r \le \sqrt{2}$, we have

$$F_R(r) = \sqrt{r^2-1} + \left[\frac{\sqrt{r^2-1}}{2} + \frac{r^2}{2}\sin^{-1}\left(\frac{1}{r}\right) - \frac{\sqrt{r^2-1}}{2} - \frac{r^2}{2}\sin^{-1}\left(\frac{\sqrt{r^2-1}}{r}\right)\right],$$

or

$$F_R(r) = \sqrt{r^2-1} + \frac{r^2}{2}\left[\sin^{-1}\left(\frac{1}{r}\right) - \sin^{-1}\left(\frac{\sqrt{r^2-1}}{r}\right)\right], 1 \le r \le \sqrt{2}.$$

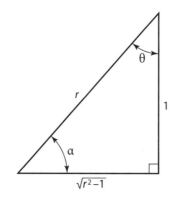

Figure 18.2.4. The geometry of a right triangle.

This result looks messy, but by making a simple observation we can make a significant simplification. Looking at Figure 18.2.4, we see that the first term in the brackets is the angle that has the sine of $1/r$ (that is, is angle α in the figure), while the second term in the brackets is the angle that has the sine of $\sqrt{r^2-1}/r$ (that is, is angle θ in the figure). Since $\alpha + \theta = \pi/2$, then, calling $\alpha - \theta = \beta$, we have $\beta = 2\alpha - \pi/2$, and so

$$F_R(r) = \sqrt{r^2-1} + \frac{r^2}{2}\left[2\sin^{-1}\left(\frac{1}{r}\right) - \frac{\pi}{2}\right], 1 \le r \le \sqrt{2}.$$

Differentiating (and then doing a little algebra) gives us the density function

$$f_R(r) = r\left[2\sin^{-1}\left(\frac{1}{r}\right) - \frac{\pi}{2}\right],$$

or, finally,

$$f_R(r) = r\left[2\csc^{-1}(r) - \frac{\pi}{2}\right], 1 \le r \le \sqrt{2}.$$

Now that we have the density function for both intervals of r, $0 \le r \le 1$ and $1 \le r \le \sqrt{2}$, we can evaluate the expectation integrals:

$$E(R) = \int_0^1 r \frac{\pi}{2} r \, dr + \left[-\int_1^{\sqrt{2}} r \frac{\pi}{2} r \, dr + 2 \int_1^{\sqrt{2}} r^2 \csc^{-1}(r) \, dr \right].$$

The first two integrals are easy, while the third can be found in integral tables:

$$\int x^2 \csc^{-1}(x) \, dx = \frac{x^3}{3} \csc^{-1}(x) + \frac{x\sqrt{x^2 - 1}}{6} + \frac{1}{6} \ln(x + \sqrt{x^2 - 1}).$$

I'll let you fill-in the (easy) details of inserting the upper and lower limits to verify that, for (a),

$$E(R) = \frac{\sqrt{2} + \ln(1 + \sqrt{2})}{3} = 0.7652,$$

the value I gave you in the problem statement.

That was pretty involved stuff, but there is a way to avoid having to calculate $f_R(r)$. We could instead, if (there's always an $if!$) we are willing to do a double integral; the payoff with that approach is that then we work directly with just the simple density functions for X and Y. That is, we'll simply integrate $\sqrt{x^2 + y^2}$ over the entire unit square, using the $joint$ probability density function of X and Y. Since X and Y are independent, then

$$f_{X,Y}(x, y) = f_X(x) f_Y(y),$$

and, since X and Y are each uniform from 0 to 1, then

$$f_{X,Y}(x, y) = \begin{array}{ll} 1, & \text{over the unit square} \\ 0, & \text{otherwise.} \end{array}$$

And so

$$E(R) = \int_0^1 \int_0^1 \sqrt{x^2 + y^2} \, dx \, dy.$$

In 1839 the great German mathematician Lejeune Dirichlet (1805–1859) wrote, "We know that the evaluation of multiple integrals generally

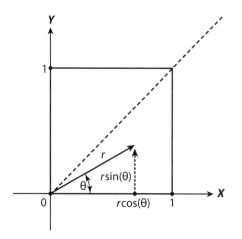

Figure 18.2.5. Switching to polar coordinates.

presents very considerable difficulties," and our double integral will indeed require us to pull a (small) rabbit out of the hat. The trick to doing the double integral is to convert to polar coordinates, as shown in Figure 18.2.5: $x = r\cos(\theta)$, $y = r\sin(\theta)$. The differential area $dxdy$ in Cartesian coordinates is the differential area $r\,dr\,d\theta$ in polar coordinates. By symmetry, all we need do is integrate over the lower diagonal half of the unit square, as the upper half will be the same; that is, we'll do the θ integration over the interval $0 \le \theta \le \pi/4$. So, doubling the integral over the lower half, we have

$$E(\boldsymbol{R}) = 2\int_0^{\pi/4}\int_0^{\sec(\theta)} r^2 \, dr d\theta.$$

where the upper limit on the inner r-integral follows from the fact that the maximum r occurs when the radius vector reaches the upper right-hand corner of the unit square, where $r/1 = \sec(\theta)$.

So, doing the easy inner integration,

$$E(\boldsymbol{R}) = \frac{2}{3}\int_0^{\pi/4} \sec^3(\theta) \, d\theta.$$

From integral tables we have

$$\int \sec^3(x)\, dx = \frac{\sec(x)\tan(x)}{2} + \frac{1}{2}\ln\left[\sec(x)+\tan(x)\right].$$

With $\tan(\pi/4) = 1$, $\sec(\pi/4) = \sqrt{2}$, $\sec(0) = 1$, and $\tan(0) = 0$, I'll let you plug in these values to verify that, for (a),

$$E(R) = \frac{\sqrt{2}+\ln\left(1+\sqrt{2}\right)}{3},$$

just as we found before.

Okay, that takes care of (a), twice. How about (b)? I'll use the double integral formulation right from the start on this part. Going from the left vertical side of the unit square to the right vertical side, as in Figure 18.2.6, we have

$$r = \sqrt{1 + (y_1 - y_2)^2}.$$

Changing variables to u and v, we have

$$E(R) = \int_0^1 \int_0^1 \sqrt{1 + (u-v)^2}\, du\, dv$$

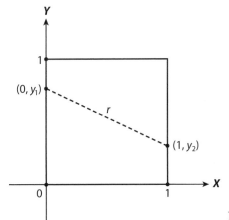

Figure 18.2.6. The geometry of (b).

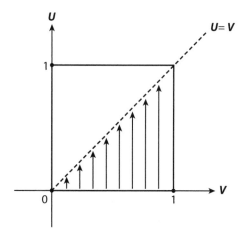

Figure 18.2.7. Altered geometry for question (b).

because the joint density function of Y_1 and Y_2 (of U and V) is 1 over the entire region of integration. Changing variables like this is conceptually a useful step because (besides avoiding having to write subscripts!) it alters our view of y_1 and y_2 (both distributed along vertical edges of the unit square) into being simply two independent, identically distributed (uniform over 0 to 1) quantities. Unlike the original geometry of Figure 18.2.6, we have the geometry of Figure 18.2.7, with U and V as along the axes of a Cartesian coordinate system.

As before, we'll take advantage of the symmetry between u and v in the integrand to replace the original integral with twice the integral over the lower half of the region of integration. That is, as Figure 18.2.7 shows, we can write

$$\int_0^1 \int_0^1 \sqrt{1 + (u-v)^2}\, du\, dv = 2\int_0^1 \left\{ \int_0^u \sqrt{1 + (u-v)^2}\, dv \right\} du.$$

Changing variables to $t = u - v$, our double integral becomes (with $dt = -dv$)

$$2\int_0^1 \left\{ \int_0^u \sqrt{1 + t^2}\, dt \right\} du.$$

The inner, t integral is (I used the Wolfram *Mathematica* Online Integrator)

$$\int \sqrt{1+t^2}\, dt = \frac{t\sqrt{1+t^2} + \sinh^{-1}(t)}{2},$$

and so our double integral is

$$\int_0^1 u\sqrt{1+u^2}\, du + \int_0^1 \sinh^{-1}(u)\, du = \left[\frac{1}{3}(u^2+1)^{3/2}\right] + [u\sinh^{-1}(u) - \sqrt{u^2+1}]$$

evaluated between the limits 0 and 1. I'll let you plug in those limits to confirm that, for (b),

$$E(\boldsymbol{R}) = -\frac{1}{3}\sqrt{2} + \frac{2}{3} + \ln(1+\sqrt{2}) = 1.0766,$$

the value I gave you in the problem statement. Since the minimum possible distance is 1, while the maximum is $\sqrt{2}$, I personally find this result nonintuitive (it seems a bit on the small side). But, there it is.

As for (c), well, *you* can set this up mathematically by writing the distance $\boldsymbol{R}$ between two random points each located anywhere in the unit square as

$$\boldsymbol{R} = \sqrt{|X_2 - X_1|^2 + |Y_2 - Y_1|^2}.$$

Let $\boldsymbol{Z} = |X_2 - X_1|$ and $\boldsymbol{W} = |Y_2 - Y_1|$, where all the X's and Y's are independent, uniformly distributed random variables from 0 to 1. Clearly, $\boldsymbol{Z}$ and $\boldsymbol{W}$ are independent, and have the same functional form for their probability densities, which I'll write as $f_{\boldsymbol{Z}}(z)$ and $f_{\boldsymbol{W}}(w)$, respectively. I'll let you work through the details, to show that

$$f_{\boldsymbol{Z}}(z) = \begin{array}{l} 2(1-z), \quad 0 \leq z \leq 1 \\ 0, \quad \text{otherwise} \end{array},$$

and

$$f_W(w) = \begin{matrix} 2\,(1-w), & 0 \le w \le 1 \\ 0, & \text{otherwise} \end{matrix}.$$

Then, since

$$R = \sqrt{Z^2 + W^2},$$

and since

$$f_{Z,W}(z,w) = f_Z(z)\,f_W(w) = 4\,(1-z)\,(1-w).$$

over the entire unit square (and zero otherwise), we have the double integral

$$E(R) = 4 \int_0^1 \int_0^1 \sqrt{z^2 + w^2}\,(1-z)(1-w)dz\,dw,$$

and this time I'll let *you* slog your way through the integrations. I will tell you that, when all the smoke clears, the answer is

$$E(R) = \frac{\sqrt{2} + 2 + 5\ln(1+\sqrt{2})}{15} = 0.5214,$$

the value I gave you in the problem statement.

18.3   COMPUTER SIMULATIONS

The calculations in the last section for the answers to (a) and (b) were long and tedious. The calculations for (c) would be characterized by the same two words (if you do them, I'm sure you'll agree!). The simulation codes for all three cases, by contrast, are short and easy. For (a) we have **adj.m** (for "adjacent sides"), for (b) we have **sts.m** (for "side-to-side"), and for (c) we have **any.m** (for "any two points"). In each case I ran the codes for a million pairs of points, and the results were:

(a) **adj.m** code $= 0.7651$ (theory $= 0.7652$)
(b) **sts.m** code $= 1.0767$ (theory $= 1.0766$)
(c) **any.m** code $= 0.5216$ (theory $= 0.5214$)

adj.m
```
total=0;
for loop=1:1000000
    r=sqrt(rand^2+rand^2);
    total=total+r;
end
total/1000000
```

sts.m
```
total=0;
for loop=1:1000000
    r=sqrt((1+(rand-rand)^2));
    total=total+r;
end
total/1000000
```

any.m
```
total=0;
for loop=1:1000000
    r=sqrt((rand-rand)^2+(rand-rand)^2);
    total=total+r;
end
total/1000000
```

Considering the agony we endured in doing the theoretical analyses, the simplicity and brevity of these codes (and their very close agreement with theory) are astounding.

Finally, here's yet another way to imagine what "the average distance on a unit square" might mean. Pick a point at random on any one of the edges (to be specific, let's say the bottom, horizontal edge), and then pick an angle at random from 0 to π radians. Then, at that angle and starting at the point, move through the square along a straight line until you reach one of the other edges (depending on the point and the angle, it could be any one of the other three edges). What's the average value of the distance traveled? Both the distribution and density functions for the distance traveled are worked out in detail in my book *Duelling Idiots* (Princeton 2012, first published in 2000), pp. 58–61 and pp. 147–155. A MATLAB® computer simulation code (**paths.m**) is also given there on pp. 244–245. The average travel distance through the unit square with this fourth interpretation is

$$\frac{1 - \sqrt{2} + 3\ln(1 + \sqrt{2})}{\pi} = 0.7098.$$

When Will the Last One Fail?

19.1 THE PROBLEM

All the stuff we use in our everyday lives eventually stops working. Eventually, *we* stop working. Who would deny that? To be specific but less personal, let's consider apparently identical toasters, coming right off the production line one after the other. Suppose we give one of these toasters to each of N people so they can make toast for breakfast each morning. Each of the N toasters therefore receives the same daily use—*but they all last different times*. One will be the first to fail, and then sometime later a second toaster will fail, and so on until the penultimate one fails, and then, alas, the Nth and last surviving toaster loses its ability to get hot.

It is an observed fact—one that can be explained theoretically if some quite reasonable physical assumptions are made, which I won't pursue here (see any good undergraduate book on probability theory)—that many different things (including toasters) fail in accordance with an exponential probability law. That is, if we write the lifetime of a toaster as T, then T is a random variable such that

$$\text{Prob}\,(T > t) = e^{-\lambda t}, t \le 0,$$

where λ is some positive constant. In particular, we then have the following two statements: $\text{Prob}\,(T > 0) = 1$ and $\lim_{t \to \infty} \text{Prob}\,(T > t) = 0$, both of which make physical sense. The first statement says a toaster won't fail before we first turn it on, and the second one says nothing lasts forever.

Now, what is λ? Obviously, it has the units of reciprocal time, since the exponent λt must be dimensionless, but that doesn't tell us what λ is. Notice that

$$\text{Prob}(T > t) = 1 - \text{Prob}(T \le t) = 1 - F_T(t)$$

where $F_T(t)$ is the probability *distribution* function of T. Since we have

$$F_T(t) = 1 - \text{Prob}(T > t) = 1 - e^{-\lambda t},$$

then the probability *density* function of T is

$$f_T(t) = \frac{d}{dt} F_T(t) = \lambda e^{-\lambda t}, t \ge 0.$$

Now, the expected value of T, that is, the average lifetime of a toaster, is given by

$$E(T) = \int_0^\infty t f_T(t)\, dt = \lambda \int_0^\infty t e^{-\lambda t} dt = \frac{1}{\lambda}.$$

In other words, λ is the reciprocal of the average lifetime. If we agree to measure time in units of the average lifetime, then with no loss in generality we can take $\lambda = 1$. So, from now on I'll write

$$F_T(t) = 1 - e^{-t}$$

and

$$f_T(t) = e^{-t}, t \ge 0.$$

One of the curious things about how a collection of seemingly identical things fail is that it can take a very long time, relative to the average failure time for an individual thing, for the last failure to occur. This is related to what is called the *memoryless property* of an exponentially distributed random variable. Here's what that means. Toasters, and any other thing that obeys an exponential law, actually seem not to fail because they gradually wear out but rather because of some sudden, out-of-the-blue event. That is, the probability a thing will survive for an addition time interval of at least b is independent of how long it has already been in use. Suppose it has been in use for a time interval of at least a; then we can express this last statement mathematically with the conditional probability (look back at Problem 17) equation

$$\text{Prob}(T > a + b | T > a) = \text{Prob}(T > b),$$

and this holds for any non-negative values for a and b.

To show that this is the case for an exponential probability law is a matter of simply calculating both sides of the conditional equation. So, on the right we have

$$\text{Prob}(T > b) = \int_b^\infty f_T(t)dt = \int_b^\infty e^{-t}dt = (-e^{-t}|_b^\infty = e^{-b}.$$

And on the left

$$\text{Prob}(T > a + b | T > a) = \frac{\text{Prob}(T > a + b, T > a)}{\text{Prob}(T > a)} = \frac{\text{Prob}(T > a + b)}{\text{Prob}(T > a)}$$

$$= \frac{e^{-(a+b)}}{e^{-a}} = e^{-b},$$

and the conditional equation is confirmed. It can be shown, in fact, that the conditional equation for the lack-of-memory property holds *only* (in the case of continuous random variables) for the exponential law.

Now, suppose that instead of toasters we imagine our things are computer chips, and that we have a vitally important security system that uses one of these chips in a central way. If the chip fails, the security system goes off-line. Thus, to improve reliability we've built N chips into the system, with initially $N - 1$ of them serving as backups. Thereafter, when the current in-use chip fails, one of the remaining backup chips is automatically switched in as a replacement. We do imagine that, even though only one chip is functional at a time, all are powered up at the same time (that is, the backup chips are "idle but hot"). This architecture motivates the first question of this problem:

(1) What's the average time until the last chip fails?

But of course we don't want to wait until the last chip fails before we plug in new backup chips to replace the failed ones; rather, when the *penultimate* (next-to-last) chip failure occurs we'll plug in new chips to replace the failed ones. So, a second question is:

(2) What's the average time until the penultimate chip failure occurs?

Calculate numerical values for both questions for $N = 2, 3, 4$, and 5. As a partial check on the numbers, for any given value of N the answer to question (2) should obviously be less than the answer to question (1).

19.2   THEORETICAL ANALYSES

Back in Problem 16 we derived the distribution functions for both the minimum and the maximum of N independent, identically distributed random variables; if U and V are the minimum and the maximum, respectively, of N random variables T_i, $1 \le i \le N$, where T_i is the lifetime of the ith chip, then

$$F_U(t) = 1 - [1 - F_T(t)]^N$$

and

$$F_V(t) = F_T^N(t).$$

The minimum and the maximum of the NT_i are, of course, the random variables for the first and the last failures, respectively. So, the densities of U and V are

$$f_U(t) = \frac{d}{dt} F_U(t) = N[1 - F_T(t)]^{N-1} f_T(t)$$

and

$$f_V(t) = \frac{d}{dt} F_V(t) = N F_T^{N-1}(t) f_T(t).$$

Or, substituting our expressions for $F_T(t)$ and $f_T(t)$ for toasters (now chips),

$$f_U(t) = Ne^{-(N-1)t}e^{-t} = Ne^{-Nt}$$

and

$$f_V(t) = N[1-e^{-t}]^{N-1}e^{-t}.$$

To answer Question (1), it is V that interests us first.
The average time until the *last* chip failure is

$$E(V) = \int_0^\infty tf_V(t)\,dt = N\int_0^\infty t[1-e^{-t}]^{N-1}e^{-t}dt,$$

an integral that looks harder to do than it is. Here's one way to evaluate
it. The binomial theorem tells us that

$$(a+b)^n = \sum_{k=0}^n \binom{n}{k}a^k b^{n-k}$$

and so, with $n = N-1$, $a = 1$, and $b = -e^{-t}$, we have

$$(1-e^{-t})^{N-1} = \sum_{k=0}^{N-1}\binom{N-1}{k}(-e^{-t})^{N-1-k}$$
$$= \sum_{k=0}^{N-1}\binom{N-1}{k}(-1)^{N-1-k}e^{-(N-1-k)t}$$
$$= \sum_{k=0}^{N-1}\binom{N-1}{k}(-1)^{N-1-k}e^{-(N-k)t}e^t.$$

So,

$$E(V) = N\int_0^\infty t\sum_{k=0}^{N-1}\binom{N-1}{k}(-1)^{N-1-k}e^{-(N-k)t}dt$$
$$= N\sum_{k=0}^{N-1}\binom{N-1}{k}(-1)^{N-1-k}\int_0^\infty te^{-(N-k)t}dt.$$

From integral tables for c, a constant, we have

$$\int te^{ct}dt = \frac{e^{ct}}{c}(t-c),$$

and so, with $c = -(N-k)$,

$$\int_0^\infty te^{-(N-k)t}dt = \left\{ \frac{e^{-(N-k)t}}{-(N-k)}\left(t + \frac{1}{N-k}\right)\right\}\Bigg|_0^\infty = \frac{1}{(N-k)^2}.$$

Thus,

$$E(V) = N\sum_{k=0}^{N-1}\binom{N-1}{k}\frac{(-1)^{N-1-k}}{(N-k)^2}.$$

This is easy to evaluate by hand for $N = 2$, 3, 4, and 5, and the result is (remember, the unit of time is the average lifetime of an individual chip):

N	$E(V)$
2	$3/2 = 1.5$
3	$11/6 = 1.833$
4	$25/12 = 2.083$
5	$137/60 = 2.283$

For Question (2), let $\mathbf{Z}$ be the random variable that represents the lifetime of the penultimate chip survivor. The distribution function of $\mathbf{Z}$ is

$$F_{\mathbf{Z}}(t) = \text{Prob}(\mathbf{Z} \le t) = \text{Prob}(all \text{ the chips have failed in time} \le t)$$
$$+ \text{Prob}(all \text{ } but \text{ } one \text{ of the chips have failed in time} \le t)$$
$$= F_T^N(t) + F_T^{N-1}(t)[1 - F_T(t)]N.$$

The second term on the right may need some explanation. The first factor is the probability that $N-1$ of the chips have failed by time t, the second factor is the probability one chip has *not* failed by time t, and the third factor is present because there are N possibilities for which chip is the last chip surviving.

Simplifying,

$$F_{\mathbf{Z}}(t) = (1-N)F_T^N(t) + NF_T^{N-1}(t).$$

Differentiating gives us the density for $\mathbf{Z}$:

$$f_{\mathbf{Z}}(t) = (1-N)NF_T^{N-1}(t)f_T(t) + N(N-1)F_T^{N-2}(t)f_T(t).$$

Before continuing with these general expressions for $F_Z(t)$ and $f_Z(t)$, notice for the special case of $N = 2$ that Z is equivalent to the *minimum* function, U. That is, when there are just two chips to begin with, then the penultimate chip that fails is the *first* chip to fail. For $N = 2$, our two general expressions for $F_Z(t)$ and $f_Z(t)$ reduce to

$$F_Z(t) = -F_T^2(t) + 2F_T(t)$$

and

$$f_Z(t) = -2F_T(t)f_T(t) + 2f_T(t).$$

Then, if you look back at the expressions for $F_U(t)$ and $f_U(t)$, for the special case of $N = 2$, you'll see that they reduce to

$$F_U(t) = 1 - [1 - F_T(t)]^2 = -F_T^2(t) + 2F_T(t)$$

and

$$f_U(t) = 2[1 - F_T(t)]f_T(t) = -2F_T(t)f_T(t) + 2f_T(t),$$

which agrees with $F_Z(t)$ and $f_Z(t)$ for $N = 2$. This agreement is reassuring as, without such agreement, we'd *know* that we had made a mistake in our equations for Z.

Now, as we did before, substitution of the formulas for $F_T(t)$ and $f_T(t)$ into the general formula for $f_Z(t)$ gives

$$f_Z(t) = (1 - N)N(1 - e^{-t})^{N-1}e^{-t} + N(N-1)(1 - e^{-t})^{N-2}e^{-t}$$

or,

$$f_Z(t) = N(N-1)(1 - e^{-t})^{N-2}e^{-2t}.$$

So, the formal answer to Question (2) is

$$E(Z) = \int_0^\infty t f_Z(t)\,dt = N(N-1)\int_0^\infty t[1 - e^{-t}]^{N-2}e^{-2t}\,dt.$$

Using the binomial theorem as before,

$$(1 - e^{-t})^{N-2} = \sum_{k=0}^{N-2} \binom{N-2}{k} (-e^{-t})^{N-2-k}$$

$$= \sum_{k=0}^{N-2} \binom{N-2}{k} (-1)^{N-2-k} e^{-(N-2-k)t}$$

$$= \sum_{k=0}^{N-2} \binom{N-2}{k} (-1)^{N-2-k} e^{-(N-k)t} e^{2t}.$$

So,

$$E(\mathbf{Z}) = N(N-1) \int_0^\infty t \sum_{k=0}^{N-2} \binom{N-2}{k} (-1)^{N-2-k} e^{-(N-k)t} dt$$

$$= N(N-1) \sum_{k=0}^{N-2} \binom{N-2}{k} (-1)^{N-2-k} \int_0^\infty t e^{-(N-k)t} dt.$$

But this last integral is the very same integral we got before, and so we immediately have

$$E(\mathbf{Z}) = N(N-1) \sum_{k=0}^{N-2} \binom{N-2}{k} \frac{(-1)^{N-2-k}}{(N-k)^2}.$$

This is just as easy to evaluate by hand as was $E(V)$, and we quickly get:

N	$E(\mathbf{Z})$
2	$1/2 = 0.5$
3	$5/6 = 0.833$
4	$13/12 = 1.083$
5	$77/60 = 1.283$

As we physically expected, $E(\mathbf{Z}) < E(V)$ for a given N. But the really striking relation that our two tables show is that $E(\mathbf{Z})$ is less than $E(V)$ by (it seems) *exactly* 1, *always*, independent of N. Can you prove this? In view, however, of the earlier discussion of the memoryless property of exponential random variables, do these numerical results now seem obvious?

Who's Ahead?

20.1 THE PROBLEM

Suppose two candidates for office, P and Q, receive p and q votes, respectively, where $p > q$. That is, P wins. That final result is definitively established, however, only after the ballot-counting process is completed. During the counting process the lead can switch back and forth, depending on the particular order in which the individual ballots are processed. A famous result in nineteenth-century probability theory called the *ballot theorem* says that the probability P is *always* ahead of Q from the first counted ballot is given by

$$\frac{p-q}{p+q}.$$

This simple expression has a rather surprising implication. Suppose, for example, that P wins by a 4-to-1 margin (a huge victory), with $p = 400$ and $q = 100$. Then the ballot theorem says P is always ahead of Q in the count with probability

$$\frac{400-100}{400+100} = \frac{300}{500} = 0.6,$$

which says that, even with P's huge victory, the probability is a not insignificant 0.4 that sometime during the counting process Q will be in at least a tie with P.

The ballot theorem was first proved by the English mathematician William Allen Whitworth (1840–1905) in 1878, although many modern writers still (mistakenly) attribute it to the French mathematician Joseph Bertrand (1822–1900), who indeed also derived the result, but not until nearly a decade later, in 1887. The result is so simple in appearance that it seems as though it should have an equally simple derivation. It does, but it's simple only in the math required. What makes the math simple is an extremely clever geometrical observation. Before I show it to you, try your own hand at proving the ballot theorem. You'll then better appreciate the trick!

20.2   THEORETICAL ANALYSIS

As the ballots are taken from the voting box to be counted, we can plot the results as shown in Figure 20.2.1, where the horizontal axis is the number of ballots processed so far and the vertical axis is the current net total count for P. That is, the positive vertical axis means P is ahead (Q is behind), the negative vertical axis means P is behind (Q is ahead), and the horizontal axis itself represents a tie. The plot (called a *path*) shown in Figure 20.2.1 is for the particular vote-counting sequence that starts with *p p p q p q q p q q*. . . . We'll call any path that has this particular start (no matter what happens thereafter as additional votes are counted) a *bad* path since they all are paths in which P is not always ahead of Q (because we have a tie after the tenth vote is counted). A *good* path would be any path that is always above the horizontal axis after the first vote is counted. Path geometry seems simple enough, but it is powerful enough to be the key to solving our problem.

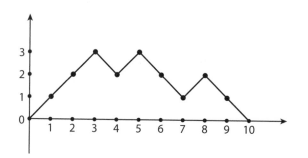

Figure 20.2.1. A bad counting path.

With our observation in mind, we can now conclude that *all* paths that begin with a vote for Q are bad paths, since they immediately fall below the horizontal axis. The probability of the first counted ballot being for Q is

$$\frac{q}{p+q}.$$

Next, consider the case where P wins the election with the final count of 4 to 3. Figure 20.2.2 shows the path for the particular counting sequence $p\,p\,q\,q\,q\,p\,p$. This path starts off looking like a good path, but eventually it "goes bad" after the fourth ballot because, at that point, there is a tie.

   Since there is a tie, then at that point we know (by definition!) there have been an equal number of votes for P and Q. This isn't as trivial as it may sound, because it means that it is possible to rearrange the counting sequence from starting as $p\,p\,q\,q$ to $q\,q\,p\,p$; that is, up to the first tie, replace each p with a q and each q with a p, as shown in Figure 20.2.3. The resulting new path is called a reflected path (up to the first tie). It is always possible to do this with a "good path gone wrong" because of the equal number of votes for P and for Q up to the first tie. Doing this has turned the good path gone wrong into a path "bad from the start," that is, into a path that we have already considered in the previous discussion.

   The reverse reflection is also possible, too, and that means there is a one-to-one correspondence between good paths gone wrong and paths

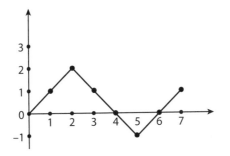

Figure 20.2.2. Another bad counting path.

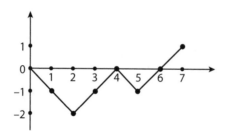

Figure 20.2.3. The reflection of the path in Figure 20.2.2.

bad from the start. That is, they have the same probability. So, the total probability of bad paths is just

$$2\frac{q}{p+q}.$$

All other paths are good paths, and so the probability of good paths, paths for which P is always ahead of Q, is

$$1 - 2\frac{q}{p+q} = \frac{p+q-2q}{p+q} = \frac{p-q}{p+q},$$

and, just like that, we have the ballot theorem.

Plum Pudding

21.1 THE PROBLEM

Problem 971 on p. 333 of the fifth edition of Whitworth's *Choice and Chance* (published in 1901, with the first edition appearing in 1867) is as follows: "If a spherical plum pudding contains n indefinitely small plums, [then show that] the expectation of the distance of the nearest one from the surface is one $(2n + 1)$th of the radius."

I was immediately attracted to this problem (as would be anyone, of course[1]), as I could see it offered a nice illustration of how the maximum of n independent random variables might occur in a "natural way" (you'll see how, in just a bit). So, out came my pen, and in just a minute or so I had my answer. Alas, it wasn't the book's answer, but rather I calculated it to be "one $(3n + 1)$th of the radius." Realizing that Whitworth was a pretty sharp fellow (this is the same Whitworth we encountered in the last problem on leads in ballot counting), my first reaction was that somewhere, I must have gone astray. But, try as I did, I couldn't find where I had screwed up.

Then I had a heretical thought: *Maybe the book is wrong.* Maybe, I thought, it was just a mistake, with some long-dead typesetter's finger hitting the 2 key instead of the 3 key, and then Whitworth not catching the typo when reading the page proofs of his book. But of course that was perhaps just a mere frantic, desperate speculation by me.

And then I had a happier thought—my quandary could be the basis for a perfect example of the value of computer simulation. The difference between the book's answer and mine is such as to be easily distinguished by a simulation. For example, if $n = 5$, then the book's answer is that the average distance from the surface of the plum near-est the surface is $1/11 = 0.09090\ldots$ of the radius, while my result was that it was $1/16 = 0.0625$ of the radius. For $n = 9$, the book's result is $1/19 = 0.05263\ldots$, compared to my $1/28 = 0.03571\ldots$. Those are significant differences.

So, in this problem we'll do the Monte Carlo simulation first, and then we'll do a theoretical analysis. It should come as no surprise for you to learn that the code agrees extremely well with my answer and not with the one in Whitworth's book (or else you wouldn't be reading any of this!). His book's answer *is* a typo. Before reading any further, see if you can derive the correct result.

21.2 COMPUTER SIMULATION

A simulation of Whitworth's problem (an approach he could only have dreamed about in 1901) is very straightforward. With no loss in general-ity I'll take a sphere with unit radius, center it on the origin of a three-dimensional x,y,z-coordinate system, and imagine it enclosed by a cube with edge length 2. To randomly place n plums in the sphere, we simply randomly place plums in the cube and work only with the ones that fall inside the sphere. That is, we'll generate triples of numbers (x,y,z) with each number uniform from -1 to 1, but keep only those triples such that $x^2 + y^2 + z^2 < 1$. We'll keep doing this until we have n points (plums) inside the sphere.

Then we'll save the value of the distance from the origin of the plum that is the furthest from the origin (and so closest to the surface) using MATLAB®'s *max* command. Then we'll repeat the whole business for, say, a total of one million times, and take the average of the saved values.

Finally, to find the average distance from the surface of the plum closest to the surface (what Whitworth actually asked for), we'll sub-tract our average distance (from the origin) from 1. The code **plums.m** does the job. When run with the values of $n = 5$ and $n = 9$, the code's estimates for the average distance from the surface was $0.062499\ldots$ and $0.03575\ldots$, respectively. Pretty good agreement with my theoreti-cal result.

plums.m

```
n=input('What is n?')
total=0;
for loop1=1:1000000
    points=zeros(1,n);
    for loop2=1:n
        keeptrying=0;
        while keeptrying==0
            x=-1+2*rand;y=-1+2*rand;z=-1+2*rand;
            d2=x^2+y^2+z^2;
            if d2<1
                keeptrying=1;
            end
        end
        points(loop2)=sqrt(d2);
    end
    total=total+max(points);
end
1-(total/loop1)
```

21.3   THEORETICAL ANALYSIS

Plums that are randomly (uniformly) distributed throughout a sphere of radius R will have a probability of appearing in any subvolume of the sphere that is directly proportional to the size of the subvolume. Thus, the probability a plum is distance r from the center of the sphere (in any direction) is given by the volume of a *thin* spherical shell of thickness $\Delta r \ll R$ and radius r, normalized to the volume of the entire sphere (normalization makes the probability the plum is somewhere in the sphere equal to 1). This probability is, therefore,

$$\frac{4\pi r^2 \, \Delta r}{4/3 \, \pi R^3} = \frac{3r^2}{R^3} \Delta r.$$

Now, let Z be the random variable representing the distance from the origin to a plum, and $f_Z(r)$ be the probability density function of Z. Then this same probability is also given by

$$f_Z(r)\Delta r,$$

and so

$$f_Z(r) = \frac{3r^2}{R^3}, 0 \le r \le R$$
$$0, \text{otherwise.}$$

The distribution function of Z is then given by

$$F_Z(r) = \text{Prob}(Z \le r) = \int_0^r F_Z(u)\, du = \frac{3}{R^3}\int_0^r u^2 du = \frac{r^3}{R^3}, 0 \le r \le R.$$

From our earlier work in Problem 16 we know that if V is the maximum of n independent random variables, each with the distribution $F_z(r)$, then the distribution of V is

$$F_V(r) = F_Z^n(r) = \frac{r^{3n}}{R^{3n}}.$$

Thus, the probability density of V is

$$f_V(r) = \frac{d}{dr}F_V(r) = \frac{3nr^{3n-1}}{R^{3n}}.$$

So, the expected value of the distance *from the origin* of the most distant plum is

$$E(V) = \int_0^R rf_V(r)\, dr = \frac{3n}{R^{3n}}\int_0^R r^{3n} dr = \left(\frac{3n}{R^{3n}}\right)\left(\frac{R^{3n+1}}{3n+1}\right) = \frac{3nR}{3n+1}.$$

Finally, the expected distance of the plum most distant from the center of the sphere, *from the surface* of the sphere (this is the plum closest to the surface), is

$$R - \frac{3nR}{3n+1} = \frac{R}{3n+1},$$

as claimed.

NOTE

1. When I told my wife about this problem one morning, she (at first) listened attentively but, after only a minute or so, her eyes began to roll back and her head dropped onto her chest. Concerned, I asked if she was okay. She replied with a loud (simulated) snore. I responded by telling her that I was never again going to tell her of my math problems, to which she tossed off a quick "Is that a promise?" But then she relented and added, "But of course I can see how a *baker* would be interested in such a thing as where the plums are." Since you are reading this I'm pretty sure you agree with me and not with my wife.

Ping-Pong, Squash, and Difference Equations

22.1 PING-PONG MATH

Everybody has played Ping-Pong at some time in life, usually at a summer camp or at school or at a friend's home. The scoring rule is simple: a point is made by the winner of each rally (it doesn't matter who serves), and the first player to reach eleven points wins the game. There is a little caveat about having to be ahead by at least two points—a score of 11 to 9 wins, but 11 to 10 doesn't—but I'll ignore that here. The first question in this puzzler is a simple one: if two players, called P and Q, have probabilities p and $1 - p = q$, respectively, of winning any given point, what is the probability that P wins the game?

We could get an estimate for this probability as a function of p by using the approach I've used a lot in this book, that is, by writing a Monte Carlo simulation. Instead, here I'll show you a completely different computer solution. It is still based on the sheer number-crunching power of a modern computer, but this alternative technique has the two advantages of speed and accuracy. Its disadvantage is that it can require some pre-coding analysis, but I think you'll find that effort worthwhile. That's because the code's calculations will be as accurate as if we had used a formula: there will be no statistical sampling errors.

To set the problem up mathematically is not difficult, and going through the analysis in detail will show us the way to attack the far more difficult question of the next section. Here's how to answer our warm-up Ping-Pong puzzle. Define the *state* of the game to be (i, j), which means that P needs i more points to win the game and Q needs j more points to win the game. If a game is in state (i, j), we'll further define $tt(i, j)$ to be the probability P wins the game *from that state* (tt for Ping-Pong's other name, *table tennis*). The answer to our warm-up question—what is the probability, as the game is about to start, that P wins the game?—is tt(11, 11).

Now, before I show you how to calculate tt(11, 11), three additional observations are required:

(1) $tt(0, j) = 1, 1 \leq j \leq 11$, because these are the probabilities that P needs zero points to win; that is, P *has won*;

(2) $tt(i, 0) = 1, 1 \leq i \leq 11$, because these are the probabilities that Q needs zero points to win; that is, Q *has won* (and so P *has lost*);

and

(3) tt(0, 0) has no meaning, as it is impossible for the game to be in a state where both P and Q have won.

Now, how do we find tt(11, 11)?

Suppose the game is in state (i, j). Then, with probability p the state changes to $(i - 1, j)$ because P wins the point, and with probability $(1 - p)$ the state changes to $(i, j - 1)$ because Q wins the point. From this we immediately have the equation

$$tt(i, j) = p\, tt(i - 1, j) + (1 - p)\, tt(i, j - 1),$$

a *double*-indexed difference equation. At this point I should tell you that there *is* a beautiful theory for handling state-transitioning problems like this, called *Markov chain theory*, after the Russian mathematician Andrei Markov (1856–1922), but it's far more powerful mathematical artillery than we need here. Instead, we'll work directly with the above difference equation to arrive at a numerical answer.

To start, think of $tt(i, j)$ as the entry in the ith row and the jth column of a matrix array (called a *probability state-transition matrix*), as shown in Figure 22.1.1. You'll notice that the first row and column ($i = 0$ and $j = 0$, respectively) have been filled in with the values from observations (1) and (2), above. In addition, tt(1, 1) has been given the value p (I'll show you why in just a moment). Now, to calculate $tt(i, j)$ directly from the difference equation, we need to refer to the previous row to get $tt(i - 1, j)$ *and* to the previous column to get $tt(i, j - 1)$. For row 1 and column 1 that means we have to look at the zeroth row and column, and alas,

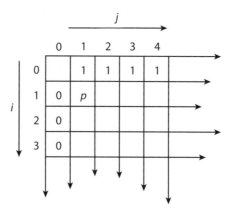

Figure 22.1.1. The probability state-transition matrix tt(i, j) for Ping-Pong.

MATLAB® doesn't support zero-indexing into a matrix. If you have software that does allow zero-indexing then you're all set to go right now, but with MATLAB® we'll have to precalculate the entries for both row 1 and column 1. Then we can use the difference equation to find the entries for all the other rows and columns in the tt state transition matrix.

To calculate row 1, set $i = 1$, and so

$$tt(1, j) = p\ tt(0, j) + (1 - p)\ tt(1, j - 1),$$

or, using observation (1),

$$tt(1, j) = p + (1 - p)\ tt(1, j - 1).$$

Thus,

$$tt(1, 1) = p + (1 - p)\ tt(1, 0) = p;$$
$$tt(1, 2) = p + (1 - p)\ tt(1, 1) = p + (1 - p)p;$$
$$tt(1, 3) = p + (1 - p)\ tt(1, 2) = p + (1 - p)p + (1 - p)^2 p;$$
$$tt(1, 4) = p + (1 - p)\ tt(1, 3) = p + (1 - p)p + (1 - p)^2 p + (1 - p)^3 p;$$

and so on. It should be obvious that, in general,

$$tt(1, j) = p[1 + (1 - p) + (1 - p)^2 + \cdots + (1 - p)^{j-1}]$$

or, recognizing the sum in the brackets as a geometric series,

$$tt(1, j) = 1 - (1 - p)^j.$$

In particular, as shown in Figure 22.1.1,

$$tt(1, 1) = 1 - (1 - p) = p.$$

To calculate column 1, set $j = 1$ and, so

$$tt(i, 1) = p\, tt(i - 1, 1) + (1 - p)\, tt(i, 0)$$

or, using observation (2),

$$tt(i, 1) = p\, tt(i - 1, 1).$$

Thus,

$$tt(1,1) = p\, tt(0, 1) = p \text{ (which we've already established)};$$
$$tt(2, 1) = p\, tt(1, 1) = p^2;$$
$$tt(3, 1) = p\, tt(2, 1) = p^3;$$

and so on. Obviously,

$$tt(i, 1) = p^i.$$

The code **pp.m** (pp for Ping-Pong) first implements these precalculations for the first row and column entries in the tt matrix, and then calculates all the remaining entries using the difference equation. The following table shows the results.

p, Probability P wins a point	tt(11, 11), Probability P wins a game
0.3	0.0264
0.35	0.0772
0.4	0.1744
0.45	0.3210

(*continued*)

(*continued*)

p, Probability P wins a point	tt(11, 11), Probability P wins a game
0.46	0.3551
0.47	0.3903
0.48	0.4264
0.49	0.4630
0.50	0.5
0.51	0.5370
0.52	0.5736
0.53	0.6097
0.54	0.6449
0.55	0.6790
0.6	0.8256
0.65	0.9228
0.7	0.9736

One quick check with these numbers adds to our confidence that the code is working correctly: the probability P wins the game for a given value of the point probability p is 1 minus the probability P wins the game for the point probability $1 - p$. This is the expected behavior, too, given the symmetry (or perhaps it's the asymmetry) between P and Q: when P wins Q loses, and vice versa. The probability of P winning a game is fairly sensitive to the value of p: if $p < 0.4$, P almost always loses, while if $p > 0.6$, P almost always wins.

```
pp.m
p=0.51;q=1-p;
for j=1:11
    t(1,j)=1-q^j;
    t(j,1)=p^j;
end
for i=2:11
    for j=2:11
        t(i,j)=p*t(i-1,j)+q*t(i,j-1);
    end
end
t(11,11)
```

22.2   SQUASH MATH IS HARDER!

Consider now a similar but much more computationally demanding problem, from the game of squash. Squash is very much like Ping-Pong in its scoring rule, with one significant difference. Our two players (still P and Q) can each score a point only by winning a rally in which they served. If the server wins the rally, then he keeps the serve; if he loses the rally, then the next serve goes to the other player. The winner of a squash game is the first player to reach nine points. For this second puzzler, we'll take P and Q as having equal ability; each has probability $1/2$ of winning any given rally (but don't forget that the rally winner gets a point only if he was the server). If the server loses a rally the service changes hand but the point totals for both players remain unchanged.

Here's our new question: how big an advantage is there for the player who gets the first service?

We can set this problem up mathematically pretty nearly the same way we did for Ping-Pong, but now we have to take into account the exchange of service. To do that, we'll define *two* state-transition matrices based on the same definition for state that we used before:

$ps(i, j)$ = probability P wins the game if P served from state (i, j)

and

$qs(i, j)$ = probability P wins the game if Q served from state (i, j).

We now have *two* double-indexed difference equations, using the same argument we used in Ping-Pong on how the game state changes as a function of who wins a rally in state (i, j):

$$ps(i, j) = \frac{1}{2}\, ps(i - 1, j) + \frac{1}{2}\, qs(i, j)$$

and

$$qs(i, j) = \frac{1}{2}\, ps(i, j) + \frac{1}{2}\, qs(i, j - 1).$$

To answer our question, all we need do is compare $ps(9, 9)$ to $qs(9, 9)$. Notice, though, that our two difference equations are cross-coupled. This problem is clearly much more complicated than was Ping-Pong.

As in the Ping-Pong analysis, we can make some preliminary observations:

(1) $ps(0, j) = 1, 1 \le j \le 9$;
(2) $qs(0, j) = 1, 1 \le j \le 9$;
(3) $ps(i, 0) = 0, 1 \le i \le 9$;
(4) $qs(i, 0) = 0, 1 \le i \le 9$;
(5) $ps(0, 0)$ and $qs(0, 0)$ have no meaning.

Figure 22.2.1 shows the structures of both the ps and the qs probability state-transition matrices, where you can see that I have used these observations to fill in the values of the zeroth row and column for each matrix,

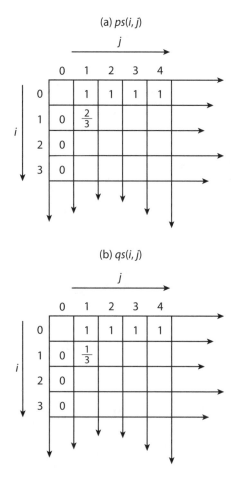

Figure 22.2.1. The two probability state-transition matrices for squash.

as well as the value of element (1, 1) for each matrix. Now, here's how I calculated the (1, 1) probabilities, and all the other entries for row 1 and column 1 in each matrix.

To calculate the values of column 1 of the ps and qs matrices, set $j = 1$. Then,

$$\text{ps}(i, 1) = \frac{1}{2}\,\text{ps}(i-1, 1) + \frac{1}{2}\,\text{qs}(i, 1)$$

and

$$\text{qs}(i, 1) = \frac{1}{2}\,\text{ps}(i, 1) + \frac{1}{2}\,\text{qs}(i, 0) = \frac{1}{2}\,\text{ps}(i, 1).$$

Thus,

$$\text{ps}(i, 1) = \frac{1}{2}\,\text{ps}(i-1, 1) + \frac{1}{2}\left[\frac{1}{2}\,\text{ps}(i,1)\right] = \frac{1}{2}\,\text{ps}(i-1, 1) + \frac{1}{4}\,\text{ps}(i, 1),$$

or,

$$\frac{3}{4}\,\text{ps}(i, 1) = \frac{1}{2}\,\text{ps}(i-1, 1),$$

and so

$$\text{ps}(i, 1) = \frac{2}{3}\,\text{ps}(i-1, 1).$$

Thus,

$$\text{ps}(1, 1) = \frac{2}{3}\,\text{ps}(0, 1) = \frac{2}{3}.$$

Also,

$$\text{qs}(1, 1) = \frac{1}{2}\,\text{ps}(1, 1) = \frac{1}{3}.$$

From $ps(i, 1) = 2/3\, ps(i-1, 1)$ we have

$$ps(2, 1) = \frac{2}{3}\, ps(1, 1) = \left(\frac{2}{3}\right)\left(\frac{2}{3}\right) = \left(\frac{2}{3}\right)^2,$$

$$ps(3, 1) = \frac{2}{3}\, ps(2, 1) = \left(\frac{2}{3}\right)^3,$$

and so on. In general, column 1 of the ps matrix is

$$ps(i, 1) = \left(\frac{2}{3}\right)^i.$$

And since $qs(i, 1) = 1/2\, ps(i, 1)$, column 1 of the qs matrix is

$$qs(i, 1) = \frac{1}{2}\left(\frac{2}{3}\right)^i.$$

To calculate row 1 of the ps and qs matrices, set $i = 1$, and so

$$ps(1, j) = \frac{1}{2}\, ps(0, j) + \frac{1}{2}\, qs(1, j) = \frac{1}{2} + \frac{1}{2}\, qs(1, j)$$

and

$$qs(1, j) = \frac{1}{2}\, ps(1, j) + \frac{1}{2}\, qs(1, j-1).$$

Thus,

$$qs(1, j) = \frac{1}{2}\left[\frac{1}{2} + \frac{1}{2}\, qs(1, j)\right] + \frac{1}{2}\, qs(1, j-1)$$

$$= \frac{1}{4} + \frac{1}{4}\, qs(1, j) + \frac{1}{2}\, qs(1, j-1),$$

or, after a couple of easy algebraic steps,

$$qs(1, j) = \frac{1}{3} + \frac{2}{3}\, qs(1, j-1).$$

So,

$$qs(1, 1) = \frac{1}{3} + \frac{2}{3}\, qs(1, 0) = \frac{1}{3}$$

which we've already established (but it's nice to see it again!),

$$qs(1, 2) = \frac{1}{3} + \frac{2}{3}\, qs(1, 1) = \frac{1}{3} + \left(\frac{2}{3}\right)\left(\frac{1}{3}\right),$$

$$qs(1, 3) = \frac{1}{3} + \frac{2}{3}\, qs(1, 2) = \frac{1}{3} + \left(\frac{2}{3}\right)\left(\frac{1}{3}\right) + \left(\frac{2}{3}\right)^2\left(\frac{1}{3}\right),$$

and so on, In general,

$$qs(1, j) = \frac{1}{3}\left[1 + \frac{2}{3} + \left(\frac{2}{3}\right)^2 + \cdots + \left(\frac{2}{3}\right)^{j-1}\right],$$

or, recognizing a geometric series, we have (with just a little algebra) that

$$qs(1, j) = 1 - \left(\frac{2}{3}\right)^j.$$

And finally, since $ps(1, j) = 1/2 + 1/2\, qs(1, j)$, we have

$$ps(1, j) = \frac{1}{2} + \frac{1}{2} - \frac{1}{2}\left(\frac{2}{3}\right)^j,$$

or,

$$ps(1, j) = 1 - \frac{1}{2}\left(\frac{2}{3}\right)^j.$$

So, this pre-coding analysis (prompted, you'll recall, by MATLAB®'s failure to allow zero indexing into matrices) has been just a bit more involved than it was for Ping-Pong! Even so, we are still not quite ready to start writing code; we have one last task to complete. When calculating an entry in either matrix, the code must use only the values of entries already calculated (I do hope this is obvious!); the difference equations for ps and qs as they stand pose a problem on this score be-cause $ps(i, j)$ uses $qs(i, j)$, and $qs(i, j)$ uses $ps(i, j)$. Each matrix uses the other in a way resembling a "pull yourself up by your bootstraps" paradox!

The way out of this chicken-and-egg conundrum is to express *one* of ps and qs in terms *only* of already calculated values and to calculate it first—and then calculate the other. Let's do this for ps. Thus,

$$ps(i, j) = \frac{1}{2} ps(i - 1, j) + \frac{1}{2}\left[\frac{1}{2} ps(i, j)\right] + \frac{1}{2} qs(i, j - 1)]$$

$$= \frac{1}{2} ps(i - 1, j) + \frac{1}{4} ps(i, j) + \frac{1}{4} qs(i, j - 1)$$

and so

$$\frac{3}{4} ps(i, j) = \frac{1}{2} ps(i - 1, j) + \frac{1}{4} qs(i, j - 1).$$

This gives us

$$ps(i, j) = \frac{2}{3} ps(i - 1, j) + \frac{1}{3} qs(i, j - 1),$$

where the right-hand side uses only previously determined values of ps and qs to calculate $ps(i, j)$. *Then* $qs(i, j)$ is calculated.

The code **squash.m** uses this last equation for $ps(i, j)$ and the original $qs(i, j)$ equation (as well as the row 1 and column 1 pre-coding results) to calculate $ps(9, 9)$ for P serving first, and for the calculation of $qs(9, 9)$ for Q serving first. When run, the code gave $ps(9, 9) = 0.534863456 \ldots$ and $qs(9, 9) = 0.465136543. \ldots$ (The fact that these two winning prob-abilities for P add to 1 is an indication that the code is working correctly, do you see *why*? Remember, in this analysis P and Q are of equal ability.) There is a definite advantage to P to get first serve—he wins noticeably more than he loses, whereas if he is the second server he loses notice-ably more than he wins. When vastly more powerful analytical methods than I've used here are applied, it can be shown that the exact answer for $ps(9, 9)$ is

$$\frac{7,674,706}{14,348,907} = 0.534863456\ldots,$$

which is precisely what **squash.m** gives us.

---

squash.m

```
r=2/3;
for i=1:9
    ps(i,1)=r^i;
    qs(i,1)=ps(i,1)/2;
    qs(1,i)=1-ps(i,1);
    ps(1,i)=1-qs(i,1);
end
for i=2:9
    for j=2:9
        ps(i,j)=r*(ps(i-1,j)+qs(i,j-1)/2);
        qs(i,j)=0.5*(ps(i,j)+qs(i,j-1));
    end
end
ps(9,9)
qs(9,9)
```

Will You Be Alive 10 Years from Now?

23.1 THE PROBLEM

As one gets older the above question takes on ever-increasing interest. The only real answer to it is something like "Who knows?," or "Maybe, but maybe not," or "Yes, God willing." Those are all a bit less than satisfying, of course; can we at least calculate the probability of being alive 10 years from now? The answer to that question is a definite *yes*, and you can do it for yourself with information you can get right off the Web. All you need is your age (which I'm sure you know) and the so-called *life-expectancy table* for your particular situation (race and gender, which I'm also sure you know).

A life-expectancy table gives, at every age, how many more years you can expect to live. (See, for example, the U.S. Social Security Administration's website, which has a Life Expectancy Calculator indexed on gender and birthdate—what statisticians call the *cohort*.) Naturally, as you get older your expected number of years yet to live continually decreases, or at least that has been the universal experience of humanity: no exceptions have ever been observed. For example, a man alive now who is age 50 might have an expected years yet to live of 30 years, but if he makes it to age 51 he would then have an expected years yet to live of only 29.5 years. A table of this sort does not directly answer our opening question, but it does implicitly contain the answer, and what I'll now show you is how to extract the probability you'll be alive 10 (or any other number you wish) years from now from such a table.

23.2   THEORETICAL ANALYSIS

Let $p(x)$ be the probability a person alive now is still alive x years from now (*now* is time zero), where $p(0) = 1$. (Our opening question is, "What is $p(10)$?") Then, $p(x + \Delta x)$ is the probability the person is alive $x + \Delta x$ years from *now*, where $\Delta x \approx 0$, and so $p(x) - p(x + \Delta x)$ is the probability the person dies in the interval x to $x + \Delta x$ years from *now*. So, as of *now* the person has x years left to live with probability $p(x) - p(x + \Delta x)$, and the average or *expected* years left to live, ϕ, is found by integrating the product $x[p(x) - p(x + \Delta x)]$ over all x. That is,

$$\phi = \int_0^\infty x[p(x) - p(x + \Delta x)]\, dx.$$

Since

$$p(x) - p(x + \Delta x) = \frac{p(x) - p(x + \Delta x)}{\Delta x} \Delta x,$$

then, as $\Delta x \to 0$ we have

$$\frac{dp}{dx} = \lim_{\Delta x \to 0} \frac{p(x + \Delta x) - p(x)}{\Delta x},$$

and so

$$\phi = \int_0^\infty x\left\{-\frac{dp}{dx}\right\} dx = -\int_0^\infty x\, dp.$$

In the well-known integration-by-parts formula,

$$\int_0^\infty u\, dv = (uv)\big|_0^\infty - \int_0^\infty v\, du,$$

let $u = x$ (and so $du = dx$) and $dv = dp$ (and so $v = p$). Then,

$$\phi = -\left[xp(x)\big|_0^\infty - \int_0^\infty p(x)\, dx\right].$$

Or, since $\lim_{x \to 0} xp(x) = 0$ because there is some *finite* value of x (certainly less than 200 years!) when $p(x) = 0$, we have the expected years left to live from *now* as

$$\phi = \int_0^\infty p(x)dx.$$

If we move forward from *now* ($x = 0$) to some later time ($x > 0$) the expected years left to live will, more generally, be $\phi(x)$, calculated just as in the above integral except for the lower limit, which is the then new *now*, x. That is, with u as a dummy variable of integration,

$$\phi(x) = \int_x^\infty p(u)du.$$

Our first integral expression for ϕ is, in fact, $\phi(0)$.

Here's what we have so far: a person alive *now* will be alive x years from *now* with probability $p(x)$ and have a life expectancy then of $\phi(x)$, or will be dead x years from *now* with probability $1 - p(x)$ and have (obviously!) a life expectancy then of zero. So, the overall life expectancy is the average of these two possibilities, given by

$$p(x)\phi(x) + [1 - p(x)]0 = p(x)\phi(x).$$

But we know the life expectancy at age x *is* $\phi(x)$, and so

$$p(x)\phi(x) = \phi(x),$$

with the surprising solution $p(x) = 1$, surprising because $p(x) = 1$ says our person is alive for all x. He is immortal! Alas, were it but so. Math is wonderful, yes, but it isn't the Fountain of Youth. This solution is mathematically correct, but it is also physical nonsense (just as, in high school algebra, when solving quadratic equations for the number of apples in a bag we would reject negative or complex solutions as physically unacceptable). Fortunately, there is another, far more realistic solution for $p(x)$.

To find it, we write $p(x)\phi(x) = \phi(x)$ in this way:

$$p(x)\phi(x) = \int_x^\infty p(u)du.$$

This is what mathematicians call an *integral equation*, with an unknown function appearing both inside and outside an integral. Such equations can be very tough to solve, often requiring powerful, advanced techniques, but we are in luck as there is a beautiful classical solution to this integral equation. What I'll show you next is how to unravel the above integral equation to find $p(x)$ as a function of $\phi(x)$ (the values of which are the entries in a life expectancy table).

Start by writing

$$\frac{1}{\phi(x)} = \frac{p(x)}{\int_x^\infty p(u)\,du}.$$

Then, define the function $P(u)$ as the indefinite integral of $p(u)$, and so

$$\frac{1}{\phi(x)} = \frac{p(x)}{P(\infty) - P(x)}.$$

(I've put this equation in bold because I'll refer back to it in just a few steps.) Integrating both sides gives

$$\int_0^x \frac{du}{\phi(u)} = \int_0^x \frac{p(u)}{P(\infty) - P(u)}\,du.$$

Next, change variable to

$$g(u) = P(\infty) - P(u),$$

and so, remembering that $p(u)$ is the derivative of $P(u)$,

$$\frac{dg}{du} = -p(u),$$

and so

$$du = -\frac{dg}{p(u)}.$$

Thus,

$$\int_0^x \frac{du}{\phi(u)} = \int_{P(\infty)-P(0)}^{P(\infty)-P(x)} \frac{p(u)}{g(u)}\left\{-\frac{dg}{p(u)}\right\} = -\int_{P(\infty)-P(0)}^{P(\infty)-P(x)} \frac{dg}{g}$$

$$= -\ln\{g(u)\}\Big|_{P(\infty)-P(0)}^{P(\infty)-P(x)} = -\ln\left\{\frac{P(\infty)-P(x)}{P(\infty)-P(0)}\right\}.$$

Now, whatever $P(\infty)$ and $P(0)$ are, they are *constants*, so let's write $c = P(\infty) - P(0)$, and thus

$$\int_0^x \frac{du}{\phi(u)} = \ln\left\{\frac{c}{P(\infty)-P(x)}\right\}.$$

Looking back at the bold equation, we see that $P(\infty) - P(x) = p(x)\phi(x)$, and so

$$\int_0^x \frac{du}{\phi(u)} = \ln\left\{\frac{c}{p(x)\phi(x)}\right\}.$$

Since $p(0) = 1$, then

$$\int_0^0 \frac{du}{\phi(u)} = 0 = \ln\left\{\frac{c}{\phi(0)}\right\},$$

which says $c = \phi(0)$. So,

$$\int_0^x \frac{du}{\phi(u)} = \ln\left\{\frac{\phi(0)}{p(x)\phi(x)}\right\} = \ln\{\phi(0)\} - \ln\{\phi(x)\} - \ln\{p(x)\},$$

or, at last (and again in bold),

$$\mathbf{\ln\{p(x)\} = \ln\{\phi(0)\} - \ln\{\phi(x)\} - \int_0^x \frac{du}{\phi(u)}.}$$

To see how to use this result in an actual calculation, suppose (just for sake of illustration) that we have the following life-expectancy table for a person of some age *now* and for the next 10 years:

Age	Life expectancy
now	$20.3 = \phi(0)$
now + 1	19.5
now + 2	18.9
now + 3	18.2
now + 4	17.6
now + 5	16.9
now + 6	16.2
now + 7	15.6
now + 8	15.0
now + 9	14.4
now + 10	$13.8 = \phi(10)$

Finally, we can answer our opening question: what is $p(10)$, the probability this person will be alive 10 years from now? From the second bold equation we have

$$\ln\{p(10)\} = \ln\{20.3\} - \ln\{13.8\} - \int_0^{10} \frac{du}{\phi(u)}$$

$$= 3.01062 - 2.62467 - \int_0^{10} \frac{du}{\phi(u)} = 0.38595 - \int_0^{10} \frac{du}{\phi(u)}.$$

To evaluate the integral I'll use a numerical technique called *Simpson's method* (after the English mathematician Thomas Simpson [1710–1761], who wrote of it in 1743), which can be remarkably accurate. Here's how it works. To estimate the value of $\int_a^b f(u)du$, divide the integration interval (a,b) into an even number of equal subintervals of width h. Then (see any good calculus textbook) we have

$$\int_a^b f(u)\, du \approx \frac{h}{3}[f(a) + 4f(a+h) + 2f(a+2h) + 4f(a+3h) + 2f(a+4h)$$
$$+ \ldots + 2f(b-2h) + 4f(b-h) + f(b)].$$

In our problem $a = 0$, $b = 10$, $h = 1$ (since we have ten subintervals, each one year wide), and $f(u) = 1/\phi(u)$. We can do the numerical work in a systematic way by arranging the arithmetic as shown in the following table:

Time	ϕ	$1/\phi$	Weighting factor	Product
now	20.3	0.04926	1	0.04926
now + 1	19.5	0.05128	4	0.20512
now + 2 .	18.9	0.05291	2	0.10582
now + 3	18.2	0.05494	4	0.21976
now + 4	17.6	0.05682	2	0.11364
now + 5	16.9	0.05917	4	0.23668
now + 6	16.2	0.06173	2	0.12346
now + 7	15.6	0.06410	4	0.25640
now + 8	15.0	0.06667	2	0.13334
now + 9	14.4	0.06944	4	0.27776
now + 10	13.8	0.07246	1	0.07246

The sum of all the products in the rightmost column is 1.7937, and, as $h/3 = 1/3$, we have

$$\int_0^{10} \frac{du}{\phi(u)} \approx 0.5979,$$

and so

$$\ln\{p(10)\} = 0.38595 - 0.5979 = -0.21195.$$

Thus, the probability our person will be alive 10 years from now is

$$p(10) = e^{-0.21195} = 0.809.$$

Simple—*when you see it.*

Chickens in Boxes

24.1 THE PROBLEM (AND SOME WARM-UPS, TOO)

This book opened with a lot of commentary (mostly unhappy, I'm afraid[1]) on Marilyn vos Savant's mathematics and, if only for symmetry reasons, it seems appropriate to end with another of her puzzles. So, the final *solved* problem in this book is from her *Parade Magazine* column of August 4, 2002, in which she printed the following question from a reader:

Suppose you're on a game show. There are four boxes in an L-shaped configuration, like this:

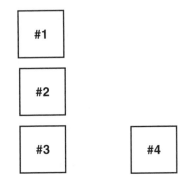

The host tells you:

(1) One of the vertical boxes contains a chicken; and
(2) one of the horizontal boxes contains a chicken.

What are the chances that a chicken is in the corner box? In one way, the chances seem to be 1 in 3; but in another way, the chances seem to be 1 in 2. They can't both be right!

To that vos Savant added a little challenge of her own, writing, "What's the answer, readers? . . . Also answer this: Which box is most likely to contain a chicken? We'll publish the first letter that correctly answers both questions."

Well, who could resist *that*? I soon had the solution in hand, but I didn't bother sending it in. I figured plenty of her other readers would. I simply stuck the solution in my files and forgot about it, recalling it only after I started writing this book. Apparently nobody else sent anything in either, however, as vos Savant never—so far as I know, and I looked pretty hard—wrote anything more on chickens in boxes.

This problem is one that can be solved in about two minutes with the honorable technique of *counting*. Don't dismiss counting as only for kids; you'll recall we used that approach in an earlier problem dealing with nontransitive bingo cards. If you can solve a good probability problem by simply counting, the very first things you should count are your blessings! There is, in fact, a large number of other interesting probability classics that are also nothing but counting puzzles. And so, before I show you the chickens-in-boxes solution, I'll illustrate (let me count the ways—five times!) the counting approach, in increasing order of subtleness.

(1) Assume the birth of a boy or a girl is equally likely, and then consider a family with two children. If you are told that the older child is a boy, what is the probability that both children are boys? If you are told that at least one of the children is a boy, what is the probability that both children are boys? The mystery of this problem is that, at first glance, it is difficult to see the distinction between the two given conditions; they appear to be providing essentially the same information. And yet the two questions have quite different answers. Here's how to see that. Make the following table showing the four possible, equally likely two-children families:

Older child	Younger child
Girl	Girl
Girl	Boy
Boy	Girl
Boy	Boy

If we are told that the older child is a boy, then we know the only possibilities are the ones in the last two rows. Since only one of those rows (the last one) has both children as boys, then the probability of both children being boys (with the given condition) is 1/2. If we are told that at least one of the children is a boy, then we know the only possibilities are the ones in the last three rows. Since only one of those rows (the last one) has both children as boys, then the probability of both children being boys (with the given condition) is 1/3.

(2) A big jar contains an unknown number of pennies. Sixty of them are taken out, given a dab of red paint, and then returned to the jar. The jar is given a vigorous shaking, and then 100 pennies are drawn at random, without replacement. If 15 of the drawn pennies have a dab of red paint on them, about how many pennies were there originally in the jar? This may seem to be simply a silly game, but substitute "lake" for jar and "fish" for penny and you have the very clever, so-called "recapture method" used to estimate the size of the fish population in a lake. Now, how does it work? Getting 15 red pennies out of the total of 60 painted ones in the jar means we drew one-fourth of the red pennies. So, we expect that the 85 unpainted pennies we drew to also be about one-fourth of the unpainted pennies. That is, there were around 340 unpainted pennies. Combined with the 60 painted pennies, a reasonable estimate is that there were originally around 400 pennies in the jar.

(3) There are four cards on the table in front of you, each with a colored side face down. Two of those faces are black and two are white. If two of the cards are picked at random and turned face up, what is the probability they will have the same color? Most people instantly answer with 1/2, but, while intuitive, that answer is wrong. Here's why. Turn the first selected card face up. It's either black or white. Suppose it's black. That means the three cards that are still face down consist of one black card and two white cards. The probability of picking the remaining black card is 1/3. If, on the other hand, the first selected card is white, that means the three cards that are still face down consist of one white card and two black cards. The probability of picking the remaining white card is 1/3. Since each of these two possibilities has probability 1/2 of occurring, our answer is $(1/2)(1/3) + (1/2)(1/3) = 1/3$.

(4) Recall the hat-matching problem I mentioned back in the preface. Suppose there are N hats in a box, where $N = 2$, 3, or 4. Find the

exact probability for, in each of these three cases, that at least one man gets his hat back. To answer this question, all we need do is write down all $N!$ permutations of the first N positive integers and then count how many have at least one instance of the jth integer in the jth position (starting at the left). For example, if $N = 2$ there are two permutations of 1,2 (that, and 2,1). The first has two matches, and the second has none. So, the probability for the $N = 2$ case is 1/2. For $N = 3$ we have six permutations and (I'll let you confirm) four of them have at least one match. So, the probability for the $N = 3$ case is 4/6 = 2/3. If you do this for $N = 4$ you'll find that of the 24 permutations there are 15 that have at least one match, and so the probability is 15/24 = 5/8. Beyond $N = 4$ this does get a bit tiresome. The exact theoretical answer for any N is $1 - 1/2! + 1/3! - \cdots \pm 1/N!$. As $N \to \infty$ (as in the example I used in the preface of a *million* men and their hats), this becomes the power series expansion for $1 - e^{-1} \approx 0.632$, the value I gave in the preface.

(5) My final counting example is an old one, dating from a problem in Lewis Carroll's 1890s book, *Pillow Problems and a Tangled Tale*. (Carroll's real name, of course, was Charles Dodgson [1832–1898], and he was the author of *Alice in Wonderland* and *Through the Looking-Glass*.) Imagine you have an urn with a single ball in it, either black or white with equal probability. Which color it is you don't know. You drop a white ball into the urn, and so now there are two balls in the urn. Then, at random, you reach in and take a ball out, which you see is white. What is the probability that the ball remaining in the urn is also white? Carroll gave the following faulty "proof" that it is 1/2 (he *knew* it was faulty, but he was just being funny). As he put it, at the start there is a white ball in the urn with probability 1/2. You then put a white ball in, and then take a white ball out. So, nothing has changed and the probability a white ball is in the urn is still 1/2. It is astonishing how many people actually think that is okay! Here's the right way to do it. Let's define the state of our problem to be a description of what is both inside and outside the urn. Initially, there are two possible, equally likely states:

Inside urn	Outside urn
w_i	w_o
b	w_o

where, in this notation, w_o is the original outside white ball and b and w_i are the possible black or white inside balls, respectively. Then, we drop w_o into the urn, which gives us two equally likely possible states:

Inside urn	Outside urn
$w_i\, w_o$	Nothing
bw_o	Nothing

Finally, we draw out a ball (which we are told is white), creating the possibility of three equally likely states:

Inside urn	Outside urn
w_i	w_o
w_o	w_i
b	w_o

Of these three equally likely possible states, two have a white ball remaining in the urn, and so the probability of a white ball remaining in the urn is 2/3.

Using this same approach, you should now be able to answer this generalization of Dodgson's problem. An urn contains, with equal probability, either n white balls or n black balls. You first drop a white ball into the urn, then randomly withdraw a ball and find that it is white. Show that the probability the original n balls in the urn were white is $(n + 1)/(n + 2)$.

The equally likely nature of the various possible states in Dodgson's problem (and its generalization) is crucial to the analysis of the problem. Sometimes, however, making a mistake about "equally likely" can be a trap for the unwary. For example, recall my mention back in the preface of Todhunter's criticism of D'Alembert. The error that tripped up D'Alembert was in the calculation of the probability of throwing a head in two tosses of a fair coin. He said it was 2/3, while a modern student of probability would instead calculate the answer to be 3/4, as follows. If we write H for heads and T for tails, then there are four possible outcomes to the two tosses: HH, HT, TH, and TT, with each possibility having equal probability (for a fair coin) of 1/4. These four possibilities are the sample points in the sample space of the experiment of tossing a coin twice. The first three of these outcomes have a head (or two), and so the probability of the event "we see a head" is the ratio of the number of favorable outcomes (3) to the total number of outcomes (4).

How did D'Alembert get 2/3? He asserted that if H appears on the first toss, then the experiment immediately ceases, and the second toss simply doesn't occur. Thus, there are only three possible outcomes: H, TH, and TT. Since there are two "favorable" outcomes out of three total outcomes, the answer is 2/3. The error in this is that the sample points in this sample space are *not* equally likely, with the first one (H) having a probability of 1/2 and each of the other two having a probability of 1/4. What D'Alembert should have written for the answer is Prob(H) + Prob(TH) = 1/2 + 1/4 = 3/4.

Let's now apply the counting approach to vos Savant's chickens-in-boxes problem.

24.2   THEORETICAL ANALYSIS

Before we apply any special constraints, there are 16 possible states for the four boxes, each of which, independently, may or may not contain a chicken. (I'll assume the "may" and "may not" are equally likely, and so each possibility has probability 1/2.) In the following table I've listed all of these states, where a 0 means an empty box and a 1 means a box with a chicken in it. Now, let's apply the given constraints. Only rows 3, 4, 5, 6, 9, and 10 satisfy the constraint that there is one chicken in the vertical boxes (Boxes 1, 2, and 3). Only rows 2, 3, 6, 7, 10, 11, 14, and 15 satisfy the constraint that there is one chicken in the horizontal boxes (Boxes 3 and 4). Only rows 3, 6, and 10 satisfy both constraints.

Looking at those three rows, we see that only one of them (row 3) has a chicken in the corner box (Box 3), and so the probability of a chicken in the corner box is 1/3. As for vos Savant's extra-credit question, to identify the box most likely to contain a chicken, in those three rows we see that Boxes 1, 2, and 3 each have a chicken in just *one* row, while Box

4 has a chicken in *two* of the three rows. So, Box 4 is the box most likely to have a chicken.

State	Box 1	Box 2	Box 3	Box 4
1	0	0	0	0
2	0	0	0	1
3	0	0	1	0
4	0	0	1	1
5	0	1	0	0
6	0	1	0	1
7	0	1	1	0
8	0	1	1	1
9	1	0	0	0
10	1	0	0	1
11	1	0	1	0
12	1	0	1	1
13	1	1	0	0
14	1	1	0	1
15	1	1	1	0
16	1	1	1	1

And with this dramatic calculation finished, so is this book of puzzles with solutions. The next puzzle, the last one in this book, comes with no solution because it remains an open challenge. If you solve it, fame is guaranteed. But be forewarned: while it is easy to understand, it has stumped all who have tried to solve it before you.

NOTE

1. This is a probability book, and so I've limited myself to commenting on vos Savant's often curious approach to probability mathematics. But I have

occasionally found her mathematical physics to be a bit shaky, as well. The very next question in her chickens-in-boxes column, for example, was from another reader asking about the reality (or not) of a triple rainbow. Vos Savant correctly said such a phenomenon does indeed exist, but then put it in the wrong part of the sky! She puts it just above the primary/secondary rainbows, when in fact a ground observer would have to *turn around and look away* from the primary/secondary rainbows to be looking in the right direction. You can find the mathematical physics behind the primary and higher-order rainbows in my book *When Least Is Best* (Princeton 2007), pp. 179–198.

Newcomb's Paradox

25.1 SOME HISTORY

In 1950 the mathematicians Merill M. Flood (1908–1991) and Melvin Dresher (1911–1992), while working at the RAND Corporation in Santa Monica, California (an air force think tank), jointly created a puzzle question in game theory that has bedeviled analysts ever since. I'll first give it to you in its best-known, nonprobabilistic form and then in the form that gives this last entry in the book its title (and in which probability makes an appearance). There is no solution section for this puzzle because, as I write, there is no known analysis that satisfies everybody. That's why it's the last problem in the book!

The best-known version of the Flood-Dresher puzzle is called the prisoner's dilemma, a name given to it by Albert W. Tucker (1905–1995), a Princeton University mathematician. Imagine that you and another person have been arrested and each of you has been charged with two crimes, one serious and the other not so serious. You've both vigorously claimed innocence, but are now being held in separate cells awaiting trial. There is no communication possible between the two of you. Just before the trial is to start, the prosecuting attorney from the DA's office shows up in your cell with the following offer.

There is sufficient circumstantial evidence to convict both of you of the not so serious charge, enough to get each of you a year in prison even if neither of you confesses. But if you confess, then the other person will be convicted of the more serious charge and get ten years in prison, and you will be set free. When you ask if the other person is getting the same offer, the answer is yes, and, further, when you ask what happens if both of you confess, the reply is that then both of you will get five years in prison. The puzzle question is now obvious: what should your decision be, to confess or not?

To help keep all the conditions clear in your mind, the following table of your various potential fates should help:

Actions	Other person confesses	Other person doesn't confess
You confess	You get 5 years in prison	You go free
You don't confess	You get 10 years in prison	You get 1 year in prison

To make your decision, you might use the following standard game theory reasoning. The other person will either confess or not. It will be one or the other, and which one it is has nothing to do with anything you can control. So, suppose he or she does confess. If you confess you get five years, and if you don't confess you get ten years. Clearly, you should confess *if he or she confesses.* But suppose your partner in crime doesn't confess. If you confess you go free, and if you don't confess you get one year. Clearly, you should confess *if your partner doesn't confess.* That is, you should confess no matter what the other person decides to do. For you to confess is said to be (in game theory lingo) the *dominant decision strategy.*

But here's the rub. The other person can obviously go through exactly the same reasoning process as you've just done, to conclude that his or her choice is also dictated by the dominant decision strategy of confessing. The end result is that you both confess and so you both get five years in prison! The paradox is that perfectly rational reasoning by each of you has resulted in a nonoptimal solution because if you both had simply kept quiet and said nothing, then you both would have gotten the much less severe sentence of one year in prison. Philosophers have argued (for decades) over whether this is really a paradox or merely "surprising," and the literature on the problem had, even years ago, grown to a point where nobody could possibly read it all in less than that ten-year prison sentence. And it continues to grow ever more voluminous even as I write.

It was while thinking about the prisoner's dilemma in 1960 that William Newcomb (1927–1999), a theoretical physicist at the Lawrence

Radiation Laboratory, now the Lawrence Livermore National Laboratory (LLNL), in California, created an even more perplexing puzzle. Newcomb's problem (now called *Newcomb's paradox*) was formulated to help him explore the prisoner's dilemma, and it is now generally believed that Newcomb's paradox is a generalization containing the prisoner's dilemma as a special case.

Curiously, Newcomb himself never published anything about his puzzle. Instead it first appeared in print in a 1969 paper by the Harvard philosopher Robert Nozick (1938–2002). The puzzle had been circulating via word-of-mouth in the academic community, but Nozick decided it needed a much wider audience. But what really brought Newcomb's puzzle worldwide fame was when it appeared in the well-known popular math essayist Martin Gardner's July 1973 "Mathematical Games" column of *Scientific American* (with a follow-up column in the March 1974 issue). So, here's Newcomb's paradox.

Imagine you are approached by an intelligent entity that has a finite but lengthy history of predicting human behavior with unfailing (so far) accuracy. It has, to date, never been wrong. You may think of this entity as (using Gardner's examples) a "superior intelligence from another planet, or a super-computer capable of probing your brain and making highly accurate predictions about your decisions." Or, if you like, you can think of the entity as God.[1] This entity makes the following presentation to you.

A week ago, the entity tells you, it predicted what you would do in the next few moments about the contents of those two mysterious boxes you've been wondering about that are sitting on a table in front of you. The boxes are labeled B1 and B2, and you can take either the contents of both boxes or the contents of box B2 only. The choice is entirely yours. B1 has a glass top, and you can *see* that the entity put $1,000 in that box. B2 has an opaque top, and you can't see what, if anything, is in it. The entity, however, tells you that it put nothing in B2 *if* last week it predicted you would take the contents of both boxes, or it put $1,000,000 in B2 *if* last week it predicted you would take the contents of only B2.

What is your decision? Take both boxes, or just box B2 alone? The reason why this situation is called a paradox is because there are seemingly two quite different (but each clearly rational) ways to argue about what you should do. The two ways, however, lead to opposite conclusions!

25.2   DECISION PRINCIPLES IN CONFLICT

The first line of reasoning is similar to the one we used in prisoner's dilemma, in that it is a dominance argument. As we did there, let's make a table of the various potential outcomes as a function of what you decide and what the entity predicted:

Actions	Entity predicted you'll take both boxes	Entity predicted you'll take only box B2
You take both boxes	You get $1,000	You get $1,001,000
You take only box B2	You get nothing	You get $1,000,000

Now, the entity (you reason) made its prediction a week ago and, based on that decision then, either did or didn't put $1,000,000 in B2. Whatever it did is a done deal and can't be changed by what you decide now. From the above table, it's clear that you have the dominant strategy of taking the contents of both boxes, as $1,000 is greater than nothing (the entity predicted you'd take both boxes), and $1,001,000 is greater than $1,000,000 (the entity predicted you'd take only B2).

That makes sense to a lot of people, maybe you, too. But there is another, probabilistic argument that leads to the opposite conclusion. It goes like this. We don't *know* that the entity is absolutely infallible. Yes, it's true that it hasn't been wrong yet, but its track record is finite. So, let's say it has probability p of being correct and, since it has always been right up to now, it is almost certain that p is pretty close to 1 (but we don't know that it *is* 1). So, for now it's p. Now, in decision theory there is, besides the dominant strategy principle, another equally respected principle called the *expected-utility* strategy, in which you decide what to do by maximizing the expected utility that results from your choice. The utility of an outcome is simply the product of the probability of the outcome by the value of the outcome, and the expected utility is the sum of all the individual utilities.

Suppose you decide to take both boxes. The entity would have predicted (correctly) that you would do that with probability p, and with probability $1 - p$ it would have predicted (incorrectly) that you'd take only B2. So, the expected utility resulting from the choice of taking both boxes is

$$U_{both} = 1,000p + 1,001,000(1 - p) = 1,001,000 - 1,000,000p.$$

Next, suppose you decide to take only B2. The entity would have predicted (correctly) that you would do that with probability p, and with probability $1 - p$ it would have predicted (incorrectly) that you'd take both boxes. So, the expected utility resulting from the choice of taking only B2 is

$$U_{B2} = 1,000,000p + 0(1 - p) = 1,000,000p.$$

Notice that as $p \rightarrow 1$, we have $U_{both} \rightarrow 1,000$ while $U_{B2} \rightarrow 1,000,000$.

The expected utility principle says you should decide to take only B2 *if* the entity is almost always correct. In fact, we can very loosely interpret what "almost" means since as long as $p > 0.5005$ (the entity simply flips an almost fair coin to make its prediction!) we have $U_{B2} > U_{both}$, and the expected utility principle says you should take only B2.

I think you can now clearly see the paradox in Newcomb's paradox. Two valid arguments, each eminent examples of rational reasoning, have led to exactly opposite conclusions. As Professor Nozick wrote in his 1969 paper,

> I have put this problem to a large number of people, both friends and students in class. To almost everyone it is perfectly clear and obvious what should be done. The difficulty is that these people seem to divide almost evenly on the problem with large numbers thinking that the opposite half is just being silly. Given two such compelling, opposing arguments, it will not do to rest content with one's belief that one knows what to do. Nor will it do to just repeat one of the arguments loudly and slowly. One must also disarm the opposing argument; explain away its force while showing it due respect.

Well, logicians, philosophers, mathematicians, physicists, and just plain folks have been trying to do that over the more than forty years since Nozick wrote, and the noise and confusion continue to this day.

What, you might wonder, did the creator of this puzzle think should be the choice? In a recent contribution,[2] the physicist Gregory Benford (who once shared an office with Newcomb at LLNL and often discussed the problem with him, long before it became famous) revealed that when he asked Newcomb that very question, the reply was a resigned "I would just take B2; why fight a God-like being?" I read that as meaning Newcomb, too, was as stumped by his own puzzle as everyone else!

This intellectual conundrum reminded Martin Gardner of one of the amusing little poetic jottings of the Danish scientist Piet Hein (1905–1996):

> A bit beyond perception's reach
> I sometimes believe I see
> That life is two locked boxes, each
> Containing the other's key.

And what could be a better note than that on which to end a book of probability puzzles?

NOTES

1. If you think a problem that asks you to accept the possibility of such a mysterious entity is simply silly, that it poses a situation nobody with any serious intent would suggest (outside of theology, of course), you are wrong. In a brilliantly original book, the political scientist Steven Brams used two-person game theory to study the outcomes of an ordinary person interacting with an "opponent" that possess the attributes of omniscience, omnipotence, immortality, and incomprehensibility. That is, he studied the interactions of a human "playing against" what he called a "superior being"—or, in a theological setting, against God. See his book, *Superior Beings: If They Exist, How Would We Know?* (Springer-Verlag 1983).

2. David H. Wolpert and Gregory Benford, "The Lesson of Newcomb's Paradox," *Synthese* (Online First), March 16, 2011. There are a lot of references in this paper to the vast literature on the problem.

Challenge Problem Solutions

(1) Let x and y be the lengths of the two randomly determined sides. That is, $0 < x < 1$ and $0 < y < 1$. Also, by the triangle inequality (a fancy way of saying that the shortest path between two points is along a straight line), we have $x + y > 1$. If we plot this inequality on the unit square (defining the triangular region above the diagonal from Y to X in Figure S1), then it is clear that all possible triangles (obtuse and acute) are associated with just the points in the upper diagonal triangle XYZ. That is, XYZ is the sample space of the problem, with area $1/2$. For the triangle to be obtuse, we further require that $x^2 + y^2 < 1$, or $y < \sqrt{1-x^2}$, a condition that associates the points in the shaded region in Figure S1 with obtuse triangles. The area of the shaded region is

$$\int_0^1 \sqrt{1-x^2}\, dx - \frac{1}{2},$$

where the first term (the integral) is the area of the *entire* portion of the unit square below the curve $y = \sqrt{1-x^2}$ and the second term is the area of the lower diagonal half of the unit square (which isn't in the sample space). The probability we are after is then

$$\frac{\text{area of shaded region}}{\text{area of sample space}} = \frac{\int_0^1 \sqrt{1-x^2}\, dx - 1/2}{1/2} = 2\int_0^1 \sqrt{1-x^2}\, dx - 1$$

$$= 2\left[\frac{x\sqrt{1-x^2}}{2} + \frac{1}{2}\sin^{-1}(x)\right]_0^1 - 1 = \sin^{-1}(1) - 1 = \frac{\pi}{2} - 1 = 0.5708.$$

The code **obtuse1.m** simulates this process. It randomly assigns values to x and y from 0 to 1 and, if $x + y > 1$, then it has a triangle and so further checks for the obtuse condition. It does this until a total of ten million triangles have been generated. When I ran the code several times, I got

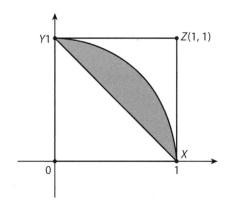

Figure S1. Sample space for obtuse triangles.

values for the probability of an obtuse triangle ranging from 0.5707 to 0.571, in fairly good agreement with theory.

```
obtuse1.m
obtuse=0;triangle=0;
while triangle<10000000
   x=rand;y=rand;
   if x+y>1
      triangle=triangle+1;
      if x^2+y^2<1
         obtuse=obtuse+1;
      end
   end
end
obtuse/10000000
```

(2) The geometric probability analysis in the *Gazette* assumes that the interior point is uniformly distributed over the interior of the entire equilateral triangle. One can certainly assume that, of course, but it isn't part of the problem statement, and to simply say "we drop the

rod and it breaks into three pieces" isn't sufficient even to imply it. One could easily imagine the rod most likely to break in its middle, with the probability of breaking elsewhere diminishing in an infinity of different ways as one moves away from the rod's midpoint. All the problem statement gives us are three values with a fixed sum (the lengths of the pieces), which means they are not even all independent. Three values do uniquely determine the location of an interior point, but uniformity over the interior of the equilateral triangle is a separate and distinct issue that needs to be established.

(3) We have two possible sets of triangle inequalities, which I'll call I and II, for $x < y$ and for $x > y$, respectively.

$$
\begin{aligned}
\text{I:} \quad & (x) + (y - x) > 1 - y \\
& (y - x) + (1 - y) > x \\
& (x) + (1 - y) > y - x
\end{aligned}
$$

$$
\begin{aligned}
\text{II:} \quad & (y) + (x - y) > 1 - x \\
& (x - y) + (1 - x) > y \\
& (y) + (1 - x) > x - y.
\end{aligned}
$$

For I, these inequalities easily reduce to

$$
\text{I:} \quad x < \frac{1}{2}, y > \frac{1}{2}, y < x + \frac{1}{2}
$$

and for II to

$$
\text{II:} \quad x > \frac{1}{2}, y < \frac{1}{2}, y > x - \frac{1}{2}.
$$

Plotting both sets of inequalities on the unit square (which is the sample space of the problem) in Figure S2 below, we see that set I gives us the upper triangular shaded area (in which x and y are such that triangles are possible from the broken pieces) and set II gives us the lower triangular shaded area (in which x and y are also such that triangles are possible from the broken pieces).

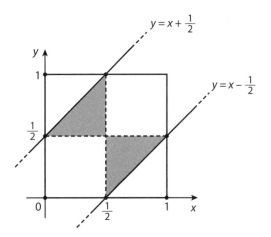

Figure S2. Sample space for broken glass rod.

Using the fundamental assumption of geometric probability (because x and y are both uniformly random), we have

$$\frac{\text{total area of shaded regions}}{\text{area of sample space}} = \frac{2\{1/2 \times 1/2 \times 1/2\}}{1} = \frac{1}{4}.$$

This is the same answer arrived at in the *Mathematical Gazette* solution to Challenge Problem 2, with its (much) less specific definition of the meaning of *breaking*. Interesting! I won't bother simulating this problem because Challenge Problem 4, using the same breaking procedure as here, starts where this problem leaves off; if something is wrong here we'll see it with the simulation of Problem 4. As you'll see, however, the Monte Carlo simulation in Problem 4 agrees *quite well* with theory.

(4) Suppose we are working with the lengths of x, $y - x$, and $1 - y$ (that is, with inequality set I from the previous problem). There can be only one obtuse angle in an obtuse triangle, and so there are just three equally likely choices for the side opposite the obtuse angle. Since the square of that side must be greater than the sum of the squares of the other two sides, we can write the following inequalities, one for each choice of side:

(a) obtuse angle opposite x: $x^2 > (y - x)^2 + (1 - y)^2$
(b) obtuse angle opposite $y - x$: $(y - x)^2 > (1 - y)^2 + x^2$
(c) obtuse angle opposite $1 - y$: $(1 - y)^2 > (y - x)^2 + x^2$.

These inequalities quickly reduce to, in order,

(a) $x > y - 1 + 1/2y$
(b) $y > 1/2(1-x)$
(c) $y < (1-2x^2)/(2(1-x))$.

All three of these inequalities are shown as shaded regions in Figure S3 (in the upper triangle, where set I applies), and if you repeat the above for set II you'll see there are three more such regions in the lower triangle where the set II inequalities apply. The total probability associated with these six regions is the probability we have an *obtuse* triangle given that we have a triangle. The area of region (b), for example, is

$$\int_0^{1/2} \left\{ \left(x + \frac{1}{2} \right) - \frac{1}{2(1-x)} \right\} dx = \frac{3}{8} - \frac{1}{2} \ln(2)$$

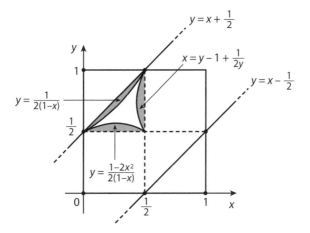

Figure S3. Sample space for alternative broken glass rod.

and (perhaps initially, with just a bit of surprise) the area of region c is the same:

$$\int_0^{1/2} \left\{ \frac{1-2x^2}{2(1-x)} - \frac{1}{2} \right\} dx = \frac{3}{8} - \frac{1}{2} \ln(2).$$

With a little thought, however, this shouldn't be surprising, as all six areas are equal by symmetry: which of the three angles is the obtuse angle is arbitrary. The only remaining issue is whether or not these areas have any overlap. In fact, they don't (for example, set $1/2(1-x) = (1-2x^2)/(2(1-x))$ and solve; you'll see this occurs *only* at $x = 0$). So, the total shaded area in the figure, representing obtuse triangles, is $6[3/8 - 1/2 \ln(2)]$, and dividing by the area of the sample space gives us the probability for an obtuse triangle (given that we have a triangle after the breaking process):

$$\frac{18/8 - 3\ln(2)}{1/4} = 9 - 12\ln(2) = 0.68223....$$

The code **obtuse2.m** simulates this process until a total of ten million triangles have been generated, checking each triangle for the obtuse condition. When run several times, the code gave estimates for the probability ranging from 0.6819528 to 0.6825122.

---

obtuse2.m

```
obtuse=0;triangle=0;
while triangle<10000000
    x=rand;y=rand;
    if x<y
        s1=x;s2=y-x;s3=1-y;
    else
        s1=y;s2=x-y;s3=1-x;
    end
    if s1<s2+s3&&s2<s1+s3&&s3<s1+s2
```

(continued)

(continued)

```
        triangle=triangle+1;
        d1=s1^2;d2=s2^2;d3=s3^2;
        if d1>d2+d3||d2>d1+d3||d3>d1+d2
            obtuse=obtuse+1;
        end
    end
end
obtuse/triangle
```

---

(5) Let X be the random variable denoting the length of the shorter piece after the first break, and so $1 - X$ represents the length of the longer piece. Let Y be the random variable denoting the length of one of the pieces into which the piece of length $1 - X$ is then broken. We now have three pieces of lengths X, Y, and $1 - X - Y$. The piece of length X is uniformly random from 0 to $1/2$ (because the *shorter* piece would in fact be the longer piece if it were longer than $1/2$), and Y is uniformly random from 0 to $1 - X$. Let x and y be particular values for X and Y, respectively. Then, from the triangle inequalities we have the following required conditions for a triangle to exist:

$$x + y > 1 - x - y, x + (1 - x - y) > y, y + (1 - x - y) > x.$$

These quickly reduce to

$$x + y > \frac{1}{2}, \frac{1}{2} > y, \frac{1}{2} > x,$$

or $y > 1/2 - x$ and $y < 1/2$. In other words, y must be in the interval $1/2 - x < y < 1/2$ for a triangle to exist. Since Y is uniform over 0 to $1 - X$, the *differential* probability dP that Y has a particular value y somewhere in the interval $1/2 - x < y < 1/2$ is

$$dP = \frac{1/2 - (1/2 - x)}{1 - x} f_X(x)\, dx,$$

where $f_X(x)$ is the probability density of X (which is 2 as X is uniform from 0 to 1/2) and $f_X(x)dx$ is the probability X has a particular value in the differential interval x to $x + dx$. So the differential probability that Y has a particular value y somewhere in the interval $1/2 - x < y < 1/2$ is

$$dP = 2\frac{x}{1-x}dx.$$

This differential probability for a triangle existing depends on X having the particular value of x: to find the *total* probability for a triangle to exist we simply integrate dP over all possible x. Thus, our probability (changing variable to $u = 1 - x$) is

$$P = 2\int_0^{1/2}\frac{x}{1-x}dx = 2\int_1^{1/2}\frac{1-u}{u}(-du) = 2\int_{1/2}^1\left(\frac{1}{u}-1\right)du$$
$$= 2\left[\ln(u)-u\right]_{1/2}^1 = 2\ln(2)-1 = 0.38629....$$

The code **long.m** simulates this breaking process ten million times and, when run several times, produced estimates for the probability P ranging from 0.3861520 to 0.3866718.

```
long.m
total=0;.
for loop=1:10000000
    x=rand;y=1-x;big=max(x,y);s=min(x,y);
    v=big*rand;u=big-v;
    m=min(u,v);l=max(u,v);
    if s+m>l&&s+l>m&&m+l>s
        total=total+1;
    end
end
total/loop
```

(6) In Figure S4 I've drawn the dartboard (a circle centered on the origin, with the equation $x^2 + y^2 = 1$). We imagine the board has been

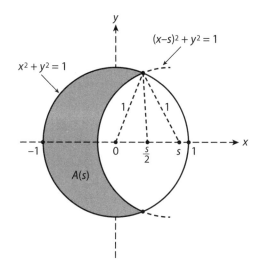

Figure S4. Geometry of the two-dart problem.

rotated after the first dart has landed so the dart is on the *x*-axis a positive distance *s* from the origin, where *s* is in the interval 0 to 1. Because
of the symmetry of the circle, we lose no generality in doing this. Next,
imagine a thin circular band of radius *s* and *very small* width Δs centered
on the origin. The probability the first dart has landed somewhere in
that band is the area of the band divided by the area of dart board (because darts are *uniform* over the dartboard): $(\pi(s + \Delta s)^2 - \pi s^2)/= 2s\Delta s$,
where we ignore all powers of Δs higher than the first (because we are
going to let $\Delta s \to 0$, that is, let $\Delta s \to ds$, in just a moment). Now, super-
imposed on the dartboard circle is another circle with unit radius, centered on the first dart, with equation $(x - s)^2 + y^2 = 1$. If the second dart
lands outside this second circle (but inside the first circle, of course, as
both darts hit the dartboard), then the second dart is at least unit distance from the first dart. That is, the second dart lands in the shaded,
lune-shaped region that has area A(*s*), where A(0) = 0. The probability
the second dart lands in the shaded region is (again) the area of the region divided by the area of the dartboard, that is, A(*s*)/π. Thus, the differential probability *dP* that the two darts are at least unit distance apart
is the product of the probabilities of these two independent events:

$$dP = \frac{2\,sA(s)\Delta s}{\pi}.$$

The total probability we are after is simply the integral of dP over all s:

$$P = \frac{2}{\pi} \int_0^1 sA(s)ds.$$

To do this integral, we first have to find $A(s)$ which is itself an integral. Writing this integral as twice the area of that portion of the lune that is above the x-axis, we have

$$A(s) = 2\int_{-1}^{s-1} \sqrt{1-x^2}\,dx + 2\int_{s-1}^{s/2}\left[\sqrt{1-x^2} - \sqrt{1-(x-s)^2}\right]dx.$$

Using the integration formula (see any good set of math tables)

$$\int \sqrt{1-x^2}\,dx = \frac{x\sqrt{1-x^2}}{2} + \frac{1}{2}\sin^{-1}(x)$$

you can show (with just a bit of algebra) that

$$A(s) = \frac{1}{2} s\sqrt{4-s^2} + 2\sin^{-1}\left(\frac{s}{2}\right).$$

Thus,

$$P = \frac{2}{\pi}\int_0^1\left[\frac{1}{2}s^2\sqrt{4-s^2} + 2s\sin^{-1}\left(\frac{s}{2}\right)\right]ds,$$

or, changing the dummy variable of integration from s to x,

$$P = \frac{1}{\pi}\int_0^1 x^2\sqrt{4-x^2}\,dx + \frac{4}{\pi}\int_0^1 x\sin^{-1}\left(\frac{x}{2}\right)dx.$$

From the integration formulas (see your math tables again),

$$\int x^2\sqrt{4-x^2}\,dx = -\frac{x(4-x^2)^{\frac{3}{2}}}{4} + \frac{4x\sqrt{4-x^2}}{8} + 2\sin^{-1}\left(\frac{x}{2}\right)$$

and

$$\int x \sin^{-1}\left(\frac{x}{2}\right) dx = \left(\frac{x^2}{2} - 1\right)\sin^{-1}\left(\frac{x}{2}\right) + \frac{x\sqrt{4 - x^2}}{4}.$$

I'll let you plug in the limits and do the arithmetic to confirm that $P = 3\sqrt{3}/4\pi = 0.41349\ldots$. The code **dd.m** (for *double-dart*) simulates this process until it has generated ten million double hits on the dartboard, where the variable *total* is the number of double hits that are at least unit distance apart. The code achieves a uniform distribution of dart landing points by imagining the circular dartboard enclosed by a square centered on the origin with edge length 2 (the diameter of the dartboard). Pairs of darts are uniformly "tossed" at the square, and used in the rest of the simulation only if *both* darts of a pair land within the circular dartboard. (This technique, called the *rejection method,* for the obvious reason, has the disadvantage of being wasteful of the random number generator. A more sophisticated way to generate points uniformly distributed over a circle is discussed in my *Digital Dice* [Princeton 2008], pp. 16–18.) When run several times, the code's estimate for P ranged from 0.4133726 to 0.4136745.

```
dd.m
hits=0;total=0;
while hits<10000000
    x1=-1+2*rand;y1=-1+2*rand;
    x2=-1+2*rand;y2=-1+2*rand;
    d1=x1^2+y1^2;d2=x2^2+y2^2;
    if d1<1&&d2<1
        hits=hits+1;
        s=(x1-x2)^2+(y1-y2)^2;
        if s>1
            total=total+1;
        end
    end
end
total/hits
```

(7) Following the hint, I'll create a simulation first and then, once you see the result, I think you will be immediately reminded of an earlier challenge problem. And that should let you see how to theoretically analyze the problem. The code **inside.m** has two distinct parts. The first is pretty straightforward, as it randomly selects three points on the circumference of a circle (a circle having, with no loss in generality, unit radius) that will be the vertices of a triangle. (Again following the hint, the first of these three random points is always the point $(1, 0)$.) The second part of the **inside.m** code is bit trickier, as it has to answer the question, "Once the code has the vertex points, how does it determine if the origin is *inside* the triangle determined by the vertex points?"

For a human, of course, this is an easy task—just look! But, lacking eyes, a computer code can't do that. There are several numerical ways to answer the question, but what I'll now show you is perhaps the simplest in both computational demands and concept. In Figure S5 you see a representative case of a triangle with vertices at $A = (X1, Y1)$, $B = (X2, Y2)$, and $C = (X3, Y3)$. The origin is shown as being inside the triangle ABC. Looking at this geometry, I think the following observation will make sense: a point P is *inside* the triangle if:

(a) P and A are on the "same side of" BC, and
(b) P and B are on the "same side of" AC, and
(c) P and C are on the "same side of" AB.

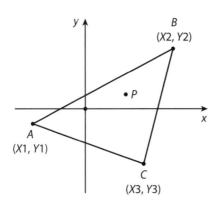

Figure S5. A point inside a triangle.

Any point in the exterior of *ABC* will not satisfy all three of these requirements. Now, here's how to understand what the phrase "same side of" means.

The equation of the line through *A* and *B* is the well-known $y = mx + b$, where the constant *m* is the slope of the line given by

$$m = \frac{Y2 - Y1}{X2 - X1}$$

and the constant *b* is the *y*-axis intercept, given by

$$b = Y1 - m\,X1.$$

Thus,

$$y = \frac{Y2 - Y1}{X2 - X1}\,x + b.$$

If we insert $x = X3$ into this equation, then we'll clearly get a *y*-value *greater* than *Y3*, just as we will if we insert the *x*-coordinate for *P*. That is, if we call the results of these two insertions *check1* and *check2*, then we'll get

$$\text{check1} = Y3 - (m\,X3 + b) < 0$$

and

$$\text{check2} = YP - (m\,XP + b) < 0,$$

where the coordinates of *P* are (*XP*, *YP*).

The importance of these calculations isn't the particular values of *check1* and *check2*, rather it is that *check1* and *check2 will have the same sign* if *P* and *C* are on the "same side of" *AB*. That means that their product will be positive if *P* and *C* are on the same side of *AB* but negative if they are on opposite sides. If *P* and *B*, and if *P* and *A*, satisfy the requirements of (b) and of (a), respectively, in the same way, then the code "knows" that *P* is inside the triangle *ABC*. When **inside.m** was run several times, each time for ten million random *ABC* triangles, the probability the origin—that is, $P = (0, 0)$—was inside *ABC* ranged from 0.2499534 to 0.2501911.

---

inside.m
```
total=0;c=2*pi;x1=1;y1=0;
for loop=1:10000000
    alpha=c*rand;beta=c*rand;
    x2=cos(alpha);y2=sin(alpha);
    x3=cos(beta);y3=sin(beta);
    m=(y2-y1)/(x2-x1);b=y1-m*x1;
    check1=y3-(m*x3+b);check2=-b;
    m=(y3-y1)/(x3-x1);b=y1-m*x1;
    check3=y2-(m*x2+b);check4=-b;
    m=(y3-y2)/(x3-x2);b=y2-m*x2;
    check5=y1-(m*x1+b);check6=-b;
    p1=check1*check2;p2=check3*check4;p3=check5*check6;
    if p1>0&&p2>0&&p3>0
        total=total+1;
    end
end
total/loop
```

Now, who could possibly look at such numerical results and not think, "I'll bet the theoretical probability is *exactly* 1/4!"? And doesn't that remind you of another challenge problem in which we also divided a finite length into three parts? Sure it does—Challenge Problem 3. In fact, Challenge Problem 7 is just a cleverly disguised version of that earlier problem. Here's why.

In Challenge Problem 3 we randomly broke a stick of unit length (in a prescribed way) into three pieces of lengths $x, y - x (= z)$, and $1 - y$, or $y, x - y (= z)$, and $1 - x$. The mathematics of that problem said that if $0 < x < 1$, $0 < y < 1$, and $0 < z < 1$, with all three lengths less than 1/2, such that Σ lengths $= 1$, then we can form a triangle, and that this happens with probability 1/4. In our current problem we are also breaking a length (the circumference of a circle) into three pieces using the same random procedure in the selection of the angles α and β (let's now suppose we measure these angles not in radians but as revolutions, with 2π radians $= 1$ revolution). Now, the three angles that separate the vertex points (if $\beta > \alpha$ these angles are α, $\alpha - \beta$, and $1 - \beta$, and if $\beta < \alpha$ these angles are β, $\alpha - \beta$, and $1 - \alpha$) satisfy the same relations as do x, y, and z (all the angles are between 0 and 1, and Σ angles $= 1$). Also, as a simple sketch or two should reveal, all the angles must be less than $1/2$ (revolution) for the origin to be inside the triangle formed by the vertex points. So, we conclude that since the mathematical description of Challenge Problem 7 is identical to that of Challenge Problem 3, the answers are identical, too.

(8) Suppose there are k teams in the league (I'll substitute $k = 6$ and $k = 10$ when we are ready to calculate numerical values). Let $p(n)$ be the probability the trophy has *not* been retired at the end of year n. Clearly, $p(1) = p(2) = 1$. There are exactly two sequences of events that result in the trophy still being unretired at the end of year n. These two sequences flow from a recognition that the champion team that year either was or was not the champion team at the end of the previous year, year $n - 1$. Let's consider each possibility separately. If the champs at the end of year n didn't win the previous year (probability $(k - 1)/k$), and the team that was the championship team that year didn't retire the trophy, we have this sequence occurring with a probability of $(k - 1)/k$ $p(n - 1)$. For the second sequence, the champs at the end of year n did win the previous year (probability $1/k$) but did not win the year before (year $n - 2$) (probability $(k - 1)/k$). The probability for this sequence is $(k - 1/k)(1/k) \, p(n - 2)$. So, we immediately have

$$p(n) = \frac{k-1}{k} p(n-1) + \frac{k-1}{k^2} p(n-2), \; p(1) = p(2) = 1.$$

This difference equation can be used to generate the numerical values of $p(n)$ for any value of n we wish, and we can keep doing that until we get to the first value of $n = n^*$ such that $p(n^*) < 1/2$. This is easy to do with a computer. For $k = 6$ we find that $n^* = 31$, because $p(30) = 0.5004$ while $p(31) = 0.4882$. For $k = 10$ we find that $n^* = 78$, because $p(77) = 0.5008$ while $p(78) = 0.4962$. To solve the difference equation analytically, try solutions of the form $p(n) = ca^n$, where c and a are constants. The difference equation will then reduce to a quadratic in a (two values), and you can use the $p(1) = p(2) = 1$ conditions to solve for the two c's that go with the two a's. The general form of the solution is $p(n) = c_1 a_1^n + c_2 a_2^n$. I warn you, however, that doing this will be a long slog through an arithmetic swamp, and by the time you are through you'll be sending your computer love notes.

(9) This problem appeared in the May 1930 issue of the British math journal *The Mathematical Gazette*. After an ingenious, geometry-only analysis, it was shown that the answer is $(4\sqrt{2} - 5)/3 = 0.2189514. \ldots$ That analysis centered on the definition of a parabola as the locus of points equidistant from a given point (the center of the green) called the *focus*, and a given line (an edge of the green) call the *directrix*. While quite clever, the analysis is also very detailed, so I decided to simply simulate the problem with the code **golf.m**. The idea is simple: if the ball is at (x, y), then the four distances to the edges of a unit square (the green) are x, $1 - x$, y, and $1 - y$, and we need only compare the minimum of them to the distance between the ball and the hole. The code does this ten million times and, with several executions, it gave estimates for the probability ranging from 0.2188581 to 0.2191731.

---

golf.m

```
total=0;
for loop=1:10000000
    x=rand;y=rand;
    V=[x,1-x,y,1-y];
    dedge=min(V);
    dcenter=sqrt((x-0.5)^2+(y-0.5)^2);
    if dcenter<dedge
        total=total+1;
    end
end
total/loop
```

A final comment: Some months after writing up this solution, I happened to come across the following in the June 2009 issue of *Mathematics*

Magazine (pp. 228–229), the solution to a problem posed in the magazine a year earlier: "A point is selected at random from the region of a regular *n*-gon. What is the probability that the point is closer to the center of the *n*-gon than it is to the *n*-gon itself?" I immediately recognized this as a generalization of the 1930 problem, which is obviously the $n = 4$ special case. The 2009 solution does require the evaluation of an integral (the 1930 solution was pure geometry), but overall, the 2009 solution for *all n* is quite pretty. The answer is $1/12\,[4 - \sec^4(\pi/2n)]$ and, if you substitute $n = 4$ into it, you do indeed get $0.2189514.\ldots$ Notice that if $n \to \infty$ (the *n*-gon approaches a circle) this probability approaches $1/3 - 1/12 = 1/4$, which is clearly correct as that is the probability the golf ball lands inside the circle (centered on the hole) that has radius one-half of the radius of the now circular green. If the green is triangular ($n = 3$), the probability is $5/27 = 0.185185.\ldots$

(10) The code **black.m** (which calls the subroutine function **draw.m**) simulates the process. When run for the various arbitrary values of b and w given in the table below (10,000 times for each row, repeated five times, with the low and high estimates given), you'll notice that independent of b and w, the probability that the last ball is black remains remarkably constant, and indeed the table strongly suggests that the exact result is *always* 1/2. A theoretical proof of this conjecture (3 pages long!), involving a double-indexed difference equation, can be found in the paper by B. F. Oakley and R. L. Perry, "A Sampling Process" (*Mathematical Gazette*, February 1965), pp. 42–44.

b	w	Probability (low/high)
1	1	0.4971–0.5073
2	3	0.4989–0.5061
3	3	0.4966–0.5080
5	4	0.4991–0.5078
7	9	0.4875–0.5055
18	13	0.4962–0.5126
15	39	0.4896–0.5037

Here's a brief walkthrough of the code. The main program, **black.m**, controls the 10,000 loops for given starting values of b (the initial number of black balls) and w (the initial number of white balls). Prior to starting the first loop, the variable *total* is set to 0; the value of *total* is the number of loops (to date) that have resulted in a black ball as the last ball in the urn. Then, in each loop, as long as there are still balls in the urn (as determined by the *while* loop), the code "goes off to" the function **draw.m**. That function accepts two input arguments, b and w, and once it has performed the drawing process it returns two output arguments to **black.m**, namely, *blackball* and *whiteball*. Those two arguments are the new number of black and white balls, respectively, remaining in the urn, and so the values of b and w are immediately updated. Then, if b has reached 0 (there are no black balls left in the urn), the code sets w to 0, which forces the *while* loop to terminate and a new execution to begin with the original values of b and w. If, on the other hand, **draw.m** has returned *whiteball* $= 0$ (and so $w = 0$), then there are only black balls left in the urn, and so *total* is incremented by 1 because the last ball then *has* to be black (and b is set to 0 to again terminate the *while* loop). Finally, if both b and w are not 0, the *while* loop is executed again (and so **draw.m** is again called). Eventually, all 10,000 loops are completed and the last line of **black.m** prints the code's estimate for the probability that the last ball in the urn is black.

---

black.m

```
total=0;
for loop=1:10000
    b=15;w=39;
    while b+w>0
        [blackball,whiteball]=draw(b,w);
        b=blackball;w=whiteball;
        if b==0
            w=0;
        elseif w==0
            total=total+1;
            b=0;
        end
    end
end
total//loop
```

The operation of **draw.m** is as follows. It starts with the reception of the current values of b and w as its input arguments from **black.m**. The drawing of the first ball is randomly made by comparing the value of *rand* (a number returned by the random number generator) to the current probability for a black ball: if the ball is black then the variable *fcolor* (for "first color") is set to 1 and b is decremented by 1 (that is, the first ball is discarded), or *fcolor* is set to 0 for white and w is decremented by 1. A *while* loop is then entered (controlled by the variable *keepgoing*, which is initially set to 1) that successively draws (and discards) balls as long as they match the color of the first ball. Now, one of two things will eventually occur: (1) a color mismatch happens or (2) all the balls matching the first color are exhausted. If a color mismatch occurs, then the control variable *keepgoing* is reset to 0 to terminate the *while* loop. If a color exhaustion occurs (see the '*if b==0*' or '*if w==0*' loops), then *keepgoing* is also reset to 0 to terminate the *while* loop. The output arguments *blackball* and *whiteball* are then set equal to the present values of b and w, respectively, and program execution is returned to **black.m**.

---

draw.m

```
function[blackball,whiteball]=draw(b,w)
   if rand<b/(w+b)
      fcolor=1;b=b-1;
   else
      fcolor=0;w=w-1;
   end
   keepgoing=1;
   while keepgoing==1;
      if fcolor==1
         if rand<b/(w+b)
            b=b-1;
            if b==0
               keepgoing=0;
            end
         else
            keepgoing=0;
         end
```

(*continued*)

(draw.m, *continued)*

```
        else
            if rand<w/(w+b)
                w=w-1;
                if w==0
                    keepgoing=0;
                end
            else
                keepgoing=0;
            end
        end
    end
    blackball=b;whiteball=w;
end
```

---

(11) For A to win on the kth turn, he must not have tossed an ace on his first $k-1$ turns (and neither did B). The probability neither player tossed an ace on the first turn (one toss each) is $(5/6)^2$, the probability neither player tossed an ace on the second turn (two tosses each) is $(5/6)^4, \ldots$, and the probability neither player tossed an ace on the $k-1$th turn ($k-1$ tosses each) is $(5/6)^{2k-2}$. So, the probability that the players arrive at the start of the kth turn both aceless is

$$\left(\frac{5}{6}\right)^2\left(\frac{5}{6}\right)^4\cdots\left(\frac{5}{6}\right)^{2k-2} = \left(\frac{5}{6}\right)^{2+4+\cdots+2k-2} = \left(\frac{5}{6}\right)^{2(1+2+\cdots+k-1)} = \left(\frac{5}{6}\right)^{k(k-1)}.$$

For A to toss an ace during the kth turn (that is, on the first toss, *or* on the second toss, *or* on the third toss, *or* . . .) is

$$\frac{1}{6}+\left(\frac{5}{6}\right)\frac{1}{6}+\left(\frac{5}{6}\right)^2\frac{1}{6}+\cdots+\left(\frac{5}{6}\right)^{k-1}\frac{1}{6} = \frac{1}{6}\left[1+\left(\frac{5}{6}\right)+\left(\frac{5}{6}\right)^2+\cdots+\left(\frac{5}{6}\right)^{k-1}\right] = 1-\left(\frac{5}{6}\right)^k.$$

So, summing over all possible k, we have the probability that A wins, $P(A)$, as

$$P(A) = \sum_{k=1}^{\infty}\left(\frac{5}{6}\right)^{k(k-1)}\left[1-\left(\frac{5}{6}\right)^k\right],$$

which is easily coded in MATLAB® to give $P(A) = 0.596794. \ldots$ The Monte Carlo code **jb.m** (jb for "James Bernoulli") simulates this game one million times, keeping track in the variable W of the number of times A wins. The logic of the simulation is pretty simple. At the start of each game, the variables A and B are each set to 0, and the variable *turn* is set to 1. Then, A tosses the die *turn* times and, if one or more aces occur, then A is set to 1; then B tosses the die *turn* times and, if one or more aces occur, then B is set to 1. This is, of course, not the way two humans would play, but the next step accounts for that. Once Player B is done, the code checks A and B. If $A = 1$, then, *irrespective* of the value of B, Player A won, W is incremented by 1, and a new game is started. If $A = 0$ and $B = 1$, then Player B won, and so W is unchanged, and a new game is started. (Look at the variable *keepgoing* that controls the *while* loop.) If $A = 0$ and $B = 0$, then neither player tossed an ace, and so *turn* is incremented by 1, and another round of tosses occurs. When run several times, **jb.m** produced estimates for $P(A)$ ranging from 0.595163 to 0.596921, nicely bracketing the theoretical value.

---

jb.m
```
W=0;p=1/6;
for loop =1:1000000
    A=0;B=0;turn=1;keepgoing=1;
    while keepgoing==1
        for loopa=1:turn
            if rand<p
                A=1;
            end
        end
        for loopb=1:turn
            if rand<p
                B=1;
            end
        end
        if A==0
            if B==0
                turn=turn+1;
```

(continued)

(jb.m, *continued*)

```
        else
            keepgoing=0;
        end
    else
        W=W+1;keepgoing=0;
    end
  end
end
W/loop
```

---

(12) Since $A^{2/3} > 0$ and $B^{2/3} > 0$ for all A and B from -1 to 1, we can think of this problem as asking for the probability that $X = A^{2/3}$ and $Y = B^{2/3}$ are such that $X + Y < 1$, where X and Y are independently distributed (each from 0 to 1). X and Y are, however, *not* uniformly distributed as are A and B. Let's write the probability density functions of X and Y, respectively, as $f_X(x)$ and $f_Y(y)$. Since X and Y are independent, their joint pdf is $f_{X,Y}(x, y) = f_X(x) f_Y(y)$. Let $Z = X + Y$. We want to compute $\mathrm{Prob}(Z < 1) = \mathrm{Prob}(X + Y < 1) = \mathrm{Prob}(Y < 1 - X)$. This probability is the probability of the shaded region in Figure S6 (the lower left diagonal half of the unit square), which we calculate by integrating the joint pdf of X and Y over that region, that is,

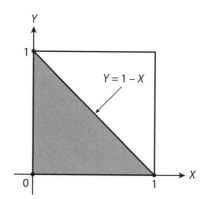

Figure S6. Sample space for Challenge Problem 12.

$$\text{Prob}\,(X + Y < 1) = \int_0^1 \int_0^{1-x} f_{X,Y}(x, y)\, dy\, dx$$

$$= \int_0^1 \int_0^{1-x} f_X(x)\, f_Y(y)\, dy\, dx$$

$$= \int_0^1 f_X(x) \left\{ \int_0^{1-x} f_Y(y)\, dy \right\} dx.$$

To do these integrals we need to find $f_X(x)$ and $f_Y(y)$. Once we find $f_X(x)$ we'll of course also know $f_Y(y)$ since X and Y are identically distributed. To find $f_X(x)$, we first calculate the distribution function of X, $F_X(x)$, and then differentiate it. So,

$$F_X(x) = \text{Prob}(X \le x) = \text{Prob}(A^{2/3} \le x)$$

$$= \text{Prob}(-x^{3/2} \le A \le x^{3/2})$$

$$= \frac{2x^{3/2}}{2} = x^{3/2}$$

because A is uniform from -1 to 1. Thus,

$$f_X(x) = \frac{d}{dx} F_X(x) = \frac{3}{2} x^{1/2}.$$

Similarly,

$$f_Y(y) = \frac{3}{2} y^{1/2}.$$

So,

$$\text{Prob}(X + Y < 1) = \int_0^1 \int_0^{1-x} \frac{9}{4} x^{1/2} y^{1/2}\, dy\, dx = \frac{9}{4} \int_0^1 x^{1/2} \left\{ \int_0^{1-x} y^{1/2}\, dy \right\} dx$$

$$\frac{9}{4} \int_0^1 x^{1/2} \left\{ \frac{2}{3} y^{3/2} \right\} \bigg|_0^{1-x} dx = \frac{3}{2} \int_0^1 x^{1/2} (1-x)^{3/2}\, dx.$$

Call the integral I, so that $\text{Prob}(X + Y < 1) = 3/2 I$. It should be obvious by inspection (but you can prove it by making the change of variable $u = 1 - x$) that

$$I = \int_0^1 x^{1/2} (1-x)^{3/2}\, dx = \int_0^1 x^{3/2} (1-x)^{1/2}\, dx,$$

and so

$$2I = \int_0^1 \left\{ x^{1/2}(1-x)^{3/2} + x^{3/2}(1-x)^{1/2} \right\} dx$$
$$= \int_0^1 x^{1/2}(1-x)^{1/2}\{1-x+x\} dx$$
$$= \int_0^1 x^{1/2}(1-x)^{1/2} dx,$$

or

$$I = \frac{1}{2}\int_0^1 \sqrt{x(1-x)}\, dx.$$

So,

$$\text{Prob}(X+Y<1) = \frac{3}{2}I = \frac{3}{4}\int_0^1 \sqrt{x(1-x)}\, dx = \frac{3}{4}\int_0^1 (x-x^2)^{1/2}\, dx.$$

Here's an elementary way to evaluate the integral. The integral is the area under the curve $y(x) = (x-x^2)^{1/2}$ as x varies from 0 to 1, and if you square this equation and do a little arranging, you'll see it becomes $(x-1/2)^2 + y^2 = (1/2)^2$. But this is simply the equation of a circle with radius $1/2$ centered on the x-axis at $x = 1/2$. So the area in question (the value of the integral) is just the upper half of the area of that circle. The area of the circle is $\pi/4$ and so the integral is $\pi/8$, and thus

$$\text{Prob}(X+Y<1) = \left(\frac{3}{4}\right)\left(\frac{\pi}{8}\right) = \frac{3\pi}{32} = 0.2945,$$

in excellent agreement with the code **final.m**.

Technical Note on MATLAB®'s Random Number Generator

Any Monte Carlo simulation code by definition uses numbers that come from some probability distribution. Some codes use a lot of random numbers; a few of the codes in this book use millions of them. MATLAB®'s random number generator provides a convenient software source for these numbers. MathWorks, the creator of MATLAB®, has gone through a series of designs for its generators, and the latest version uses a highly sophisticated combination of shift register/bit manipulation processes that, unlike earlier generators, require no multiplication or division operations. The latest MATLAB® generator, then, is very fast. At least as important, however, is that such a design results in a stupendously long period, where the *period* is the number of numbers a generator can produce before it starts repeating.

The fact that a software generator is *certain* to eventually repeat is due to the fact that the generator is what computer scientists call a *finite-state machine*. If a digital entity is built from n elementary two-state elements (that is, from n binary elements that are each either 0 or 1), then there can be at most 2^n distinct combinations of 0s and 1s (each such combination is called a *state* of the entity). To use repeating random numbers in a simulation means the numbers aren't random anymore, a situation that invalidates the underlying, fundamental assertion that the simulation somehow emulates the random process being simulated.

So, having a large period is an essential feature of a good random number generator. MATLAB®'s random number generator period is so huge that, even if producing numbers at the impressive rate of a billion per second beginning with the birth of the universe in the moment of the Big Bang, we would today have observed only an *infinitesimally tiny* fraction of the period. No simulation code you or anybody else will ever write will even begin to exhaust the ability of MATLAB® to provide the code with new random numbers.

The MATLAB® command *rand* produces a number from a distribution uniform from 0 to 1. If you want a number from a distribution uniform from a to b, where $b > a$, then $a + (b - a)$*rand* does the job. MATLAB® also provides the command *randn* that produces a number from a normal (bell-shaped) distribution with zero mean ($m = 0$) and unit standard deviation ($\sigma = 1$), that is, from a distribution with density function

$$\frac{1}{\sqrt{2\pi}} e^{-x^2/2}, \quad -\infty < x < \infty.$$

If you want a number from a normal distribution with mean m and standard deviation σ, then $m + \sigma$*randn* does the job. If you want a number from a distribution other than a uniform one or a normal one, then you'll have to write some additional code of your own: see my book *Digital Dice* (Princeton 2008), pp. 252–254, for an example on one way to do that for the case of an exponential distribution.

When you start a new MATLAB® session the generator is automatically put into a predetermined initial state; then the generator transitions from its present state to the next state as each new number is produced. The details of the state transitions are determined by the specific design of the generator, the details of which you don't need to know. This means that if you run one of the codes in this book multiple times in the same session, the first run uses one set of numbers from the generator, the second run uses a different set, and so on. That means that while the results produced by the code from one run to the next will be close (hopefully!), they won't be exactly the same.

When you are first writing a simulation code, getting different answers every time you run the code can complicate the process of debugging; is the code giving different results because its logic is in error (this is bad) or because it's getting different numbers from the generator (this is okay)? So, at the start of writing a code, it's often desirable to be able to reinitialize the generator to the same state, every time the code is run during the same MATLAB® session. This can be done by writing, somewhere in the code before the first use of the generator, the command *rand*('*state*',0).

Once the debugging process is completed, however, you'll want every new run of the code from session to session to use random numbers it has never seen before. After all, why bother running a simulation code with the same generator numbers as before—you'll just get the same answer as before! So, how do you achieve that when I've just told you that MATLAB® automatically forces the generator into the same initial state at the start of every new session? There is a simple way to do this using the *clock* command.

MATLAB®, residing in your computer, has access to your computer's knowledge of the time and uses that access to create and continually update the six-element vector *clock* with the format [year, month, day, hour, minute, second]. For example, I am typing this at one minute and a few seconds past 1 p.m. on the afternoon of February 3, 2012, and "right now" the *clock* vector is [2012, 2, 3, 13, 1, 9.5]. Since the march of time is inexorably unidirectional, we can use this vector to generate a new, unique (*almost* always) initial state every time we run a simulation code. This is done by writing, somewhere in your code before the first use of the generator, the command *rand*('*state*',100*sum(*clock*)). The command *sum* adds the six elements of the *clock* vector, and the multiplication by 100 results in an integer (for the above vector the integer is 204050) to set the initial state of the generator. A different time, a different *clock* vector, a different (usually) integer, and so a different initial state for the generator and a different stream of numbers each time you run your code.

You'll see in the codes given in this book that I haven't bothered to do anything about the initial state of the random number generator. The results I give from running the codes in this book are typical—but were produced by whatever numbers came out of the generator when I happened to run the simulations. If you run these codes on your machine (if you have MATLAB®) you'll almost certainly get results that are near mine, but not exactly the same. It is, in fact, a singular feature of Monte Carlo computer codes to get something new almost every time!

When a code runs hundreds of thousands (or even millions) of simulations and then computes the average of all those individual simulations, the results are generally ones having a lot of digits. Do all of those digits really have meaning? Almost certainly not, but nonetheless I have reported all the digits the codes in this book have generated, even though probably only the first three (or maybe four, if we're lucky) are correct. You should be skeptical of digits beyond the fourth decimal place produced by the codes in this book. If you need more digits, however, theoretical analyses of the underlying mathematics of the Monte Carlo method show that the statistical error decreases as $\sqrt{N}$, where N is the number of simulations.

For example, increasing N from 10,000 to 1,000,000 (a factor of 100) should reduce the error in the code's estimates of whatever parameters are being studied by a factor of about $\sqrt{100} = 10$. A specific illustration of this behavior is given in my book *Digital Dice*, pp. 11–15, along with an example of a quite different approach (called *variance reduction*) on pp. 223–228.

Acknowledgments

No book comes into existence just because it has an author. The author matters, of course, but lots of other people are involved, too. Over my thirty years of teaching probability theory to college engineering classes, I have had hundreds of students, all of whom have served (some even willingly!) as test subjects for problems. This is my third probability puzzle book for Princeton University Press—*Duelling Idiots* and *Digital Dice* were the first two—and the problems in all three were vetted by my students. I thank each and every one of them.

Once I deliver a raw typescript to PUP it becomes the object of intense scrutiny by many, from my editor Vickie Kearn, to her PUP colleagues Debbie Tegarden, Quinn Fusting, Dimitri Karetnikov, and Carmina Alvarez-Gaffin, to others whom I don't directly work with but I know are there. In addition, the book is transformed from delivered to finished form, one ready for the printer, by the efforts of the very important copy editor, who, for this book, was Marjorie Pannell, in Chicago. (This is the *fourth* book of mine that Marjorie has worked on.) All of these hard-working and talented folks have my sincere thanks.

Finally, my wife of fifty years, Patricia Ann, has played a very big role in all of my books, simply by understanding why I disappear for long periods of time, either to my office or to the university library, to write. And when I eventually reappear (certainly by dinnertime), there is always a hot meal and a kiss waiting for me—and sometimes the trash to take out, as well. What more could a seventy-two-year-old guy with bad eyesight whose hair (what's left of it) is turning white possibly want out of life?

Paul Nahin
Lee, New Hampshire
February 2013

Index

Also by Paul J. Nahin

Oliver Heaviside (1988, 2002)

Time Machines (1993, 1999)

The Science of Radio (1996, 2001)

An Imaginary Tale (1998, 2007)

Duelling Idiots (2000, 2002)

When Least Is Best (2004, 2007)

Dr. Euler's Fabulous Formula (2006, 2011)

Chases and Escapes (2007, 2012)

Digital Dice (2008, 2013)

Mrs. Perkins's Electric Quilt (2009)

Time Travel (2011)

Number-Crunching (2011)

The Logician and the Engineer (2013)

Searching for God in Spacetime (2014)